Social Darwinism
in American Thought

Social Darwinism in American Thought

Richard Hofstadter

WITH A NEW INTRODUCTION
BY ERIC FONER

BEACON PRESS
Boston

Beacon Press
25 Beacon Street
Boston, Massachusetts 02108-2892

Beacon Press books
are published under the auspices of
the Unitarian Universalist Association of Congregations.

First published in 1944 by University of Pennsylvania Press, Philadelphia,
and Humphrey Milford, Oxford University Press, London. First Beacon
edition, revised and reset, published by arrangement with the author in 1955.

99 98 97 96 95 94 93   8 7 6 5 4 3 2

Library of Congress Cataloging-in-Publication Data
Hofstadter, Richard, 1916–1970.
Social Darwinism in American thought / Richard Hofstadter ; with a
new introduction by Eric Foner.
p.   cm.
Originally published: Philadelphia : University of Pennsylvania
Press, 1944.
Includes bibliographical references and index.
ISBN 0-8070-5503-4 (pbk.)
1. Sociology—United States—History. 2. Social Darwinism—United
States—History. I. Title.
HM22.U5H6 1992
301'.0973—dc20   91-45525
CIP

THIS EDITION
IS FOR MY CHILDREN
DAN AND **SARAH**

Contents

INTRODUCTION BY ERIC FONER   ix

AUTHOR'S NOTE   xxix

AUTHOR'S INTRODUCTION   3

1. THE COMING OF DARWINISM   13

2. THE VOGUE OF SPENCER   31

3. WILLIAM GRAHAM SUMNER: SOCIAL DARWINIST   51

4. LESTER WARD: CRITIC   67

5. EVOLUTION, ETHICS, AND SOCIETY   85

6. THE DISSENTERS   105

7. THE CURRENT OF PRAGMATISM   123

8. TRENDS IN SOCIAL THEORY, 1890–1915   143

9. RACISM AND IMPERIALISM   170

10. CONCLUSION   201

BIBLIOGRAPHY   205

NOTES   217

INDEX   243

Introduction

Two decades have now elapsed since the untimely death of Richard Hofstadter. Despite the sweeping transformation of historical scholarship during these years, his writings continue to exert a powerful influence on how scholars and general readers alike understand the American past. Since his death, the study of political ideas—the recurring theme of Hofstadter's work—has to a considerable extent been eclipsed by the histories of family life, race relations, popular culture, and a host of other social concerns. The writings of many of his contemporaries are now all but forgotten, yet because of his penetrating intellect and sparkling literary style, Hofstadter still commands the attention of anyone who wishes to think seriously about the American past. The reissue of his first book, *Social Darwinism in American Thought*, provides an opportune moment to consider the circumstances of its composition and the reasons for its enduring influence.

Richard Hofstadter was born in 1916 in Buffalo, New York, the son of a Jewish father and a mother of German Lutheran descent. After graduating from high school in 1933, he entered the University of Buffalo, where he majored in philosophy and minored in history. As for so many others of his generation, his formative intellectual and political experience was the Great Depression. Buffalo, a major industrial center, was particularly hard hit by unemployment and social dislocation. The Depression, Hofstadter later recalled, "started me thinking about the world.... It was as clear as day that something had to change.... You had to decide, in the first instance, whether you were a Marxist or an American liberal."[1] At the university, Hofstadter gravitated toward a group of left-wing students, including the brilliant and "sometimes overpowering" (as Alfred Kazin later described her) Felice Swados, read Marx and Lenin, and joined the Young Communist League.[2]

In 1936, on the eve of his graduation, Hofstadter and Felice were married and subsequently moved to New York. Felice first worked for the National Maritime Union and International Ladies' Garment Workers Union and then took a job as a copy editor at *Time*, while Hofstadter enrolled in the graduate history program at Columbia University. Both became part of New York's broad radical political culture that centered on the Communist party in the era of the Popular Front. Hofstadter would later describe himself (with some exaggeration) as "by temperament quite conservative and timid and acquiescent,"[3] and it seems that the dynamic Felice, a committed political activist, animated their engagement with radicalism. Nonetheless, politics for Hofstadter was much more than a passing fancy; he identified himself as a Marxist and, in apartment discussions and in his correspondence with Felice's brother Harvey Swados, took part in the doctrinal debates between Communists, Trotskyists, Schachtmanites, and others that flourished in the world of New York's radical intelligentsia.

In 1938, Hofstadter joined the Communist party's unit at Columbia. The decision, taken with some reluctance (he had already startled some of his friends by concluding that the Moscow purge trials were "phony") reflected a craving for decisive action after "the hours I have spent jawing about the thing." As he explained to his brother-in-law: "I join without enthusiasm but with a sense of obligation. . . . My fundamental reason for joining is that I don't like capitalism and want to get rid of it. I am tired of talking. . . . The party is making a very profound contribution to the radicalization of the American people. . . . I prefer to go along with it now."[4]

Hofstadter, however, did not prove to be a very committed party member. He found meetings "dull" and chafed at what he considered the party's intellectual regimentation. By February 1939 he had "quietly eased myself out." His break became irreversible in September, after the announcement of the Nazi-Soviet Pact.[5] There followed a rapid and deep disillusionment—with the party (run by "glorified clerks"), with the Soviet Union ("essentially undemocratic"), and eventually with Marxism itself.[6] Yet for some years, Hofstadter continued to regard himself as a radical. "I hate capitalism and everything that goes with it," he wrote Harvey Swados soon after leaving the party. Never again, however, would he devote his energies in any sustained manner to a political cause. He became more and more preoccupied with the thought that intellectuals were unlikely to find a comfortable home in any socialist society likely to emerge in his lifetime. "People like us," he wrote, ". . . have become permanently alienated from the spirit of revolutionary movements. . . . We are not the beneficiaries of capitalism, but we will not be the beneficiaries of the socialism of the 20th century. We are the people with no place to go."[7]

Although Hofstadter abandoned active politics after 1939, his earliest work as a historian reflected his continuing intellectual engagement with radicalism. His Columbia master's thesis, written in 1938, dealt with the plight of southern sharecroppers, a contemporary problem that had become the focus of intense organizing efforts by Socialists and Communists.[8] Hofstadter showed how the benefits of New Deal agricultural policies in the cotton states flowed to large landowners, while the sharecroppers' conditions only worsened. The essay presented a devastating indictment of the Roosevelt administration for pandering to the South's undemocratic elite. Its critical evaluation of Roosevelt, a common attitude among New York radicals, would persist in Hofstadter's writings long after the political impulse that inspired the thesis had faded.

As with many others who came of age in the 1930's, Hofstadter's general intellectual approach was framed by Marxism, but in application to the American past, the iconoclastic materialism of Charles A. Beard was his greatest inspiration. "Beard was really *the* exciting influence on me," Hofstadter later remarked.[9] Beard taught that American history had been shaped by the struggle of competing economic groups, primarily farmers, industrialists, and workers. Underlying the clashing rhetoric of political leaders lay naked self-interest; the Civil War, for example, should be understood essentially as a transfer of political power from southern agrarians to northern capitalists. Differences over the tariff had more to do with its origins than with the debate over slavery. Hofstadter's first published essay, a "note" in a 1938 issue of the *American Historical Review,* took issue with Beard's emphasis on the tariff as a basic cause of the Civil War, while accepting the premise that economic self-interest lay at the root of political behavior.[10] (The homestead issue, Hofstadter argued, far outweighed the tariff as a source of sectional tension.) The article inaugurated a dialogue with the Beardian tradition that shaped much of Hofstadter's subsequent career.

While Beard devoted little attention to political ideas, seeing them as mere masks for economic self-interest, Hofstadter soon became attracted to the study of American social thought. His interest was encouraged by Merle Curti, a Marxist Columbia professor with whom Hofstadter by 1939 had formed, according to Felice, a "mutual admiration society."[11] Other than his relationship with Curti, however, Hofstadter was not particularly happy at Columbia. For three years running, he was refused financial aid. Hofstadter was gripped by a sense of unfair treatment. "The guys who got the fellowships," he complained, "are little shits who never accomplished or published anything."[12] (None of them, one can assume, had, like Hofstadter, published in the *AHR.*)

Denied financial aid, Hofstadter was forced to seek a teaching job. In the spring of 1940, he obtained a part-time position in the evening session of Brooklyn College. His first full-time job was at the downtown branch of City College, where a position opened in the spring of 1941 because of the forced departure of a professor accused of membership in the Communist party. The New York legislature's Rapp-Coudert Committee had been investigating "subversive" influences within the city colleges; eventually, some forty teachers were fired or forced to resign after being named by informants. Students initially boycotted Hofstadter's lectures as a show of support for his purged predecessor, but eventually they returned to the classroom. Ironically, Hofstadter's first full-time job resulted from the flourishing of the kind of political paranoia that he would later lament in his historical writings.

Meanwhile, having passed his comprehensive examinations, Hofstadter set out in quest of a dissertation topic. In a letter to his brother-in-law that typified Hofstadter's wry, self-deprecating sense of humor, he described the process. First, he considered writing a biography of "the old rascal Ben Wade" (the Radical Republican senator from Ohio) only to discover that Wade had destroyed most of his papers. Then he turned to Simon Cameron, Lincoln's first secretary of war, but abandoned that subject when he heard that "somebody from Indiana had been working on Cameron for 15 years." Columbia professor John A. Krout suggested a biography of Jeremiah Wadsworth, a colonial merchant who not only left abundant papers but had some admirers willing to help fund biographical research. Hofstadter, however, did not pursue the idea far—he and Felice considered Wadsworth inconsequential and kept referring to him as Jedediah Hockenpfuss. Finally, with Curti's approval, he settled on social Darwinism.[13] By mid-1940, he was hard at work, and two years later, at the precocious age of twenty-six, he completed the dissertation. *Social Darwinism in American Thought* was published by the University of Pennsylvania Press in 1944.

However serendipitous the process by which he found it, social Darwinism was the perfect subject for the young Hofstadter. It was a big topic, likely to interest a large audience, and it combined his growing interest in the history of social thought with his continuing alienation from American capitalism. It was the kind of subject, Felice wrote Harvey, "in which all his friends want to have a hand." "But in which they won't," Hofstadter added. The book focuses on the late nineteenth century and ends in 1915, the year before Hofstadter's birth. But, as he later observes, the "emotional resonances" that shaped his approach to the subject were those of his own youth, when conservatives used arguments descended from social Darwinism to justify resistance to radical political movements and government efforts to alleviate inequality. Studying social Darwinism helped explain "the disparity between our official individualism and the bitter facts of life as anyone could see them during the great depression."[14]

Social Darwinism in American Thought describes the broad impact on intellectual life of the scientific writings of Charles Darwin and the growing use of such Darwinian ideas as "natural selection," "survival of the fittest," and "the struggle for existence" to reinforce conservative, laissez-faire individualism. The book begins by tracing the conquest of Darwinian ideas among American scientists and liberal Protestant theologians, a conquest so complete that by the Gilded Age "every serious thinker felt obligated to reckon with" the implications of Darwin's writings. Hofstadter then examines the "vogue" of Herbert Spencer, the English philosopher who did more than any other individual to define nineteenth-century conservatism. Spencer, of course, preceded Darwin; well before the publication of *The Origin of Species,* Spencer not only coined the term "survival of the fittest" but developed a powerful critique of all forms of state interference with the "natural" workings of society, including regulation of business and public assistance to the poor. But Spencer's followers seized upon the authority of Darwin's work to claim scientific legitimacy for their outlook and to press home the analogy between the natural and social worlds, both of which, they claimed, evolved according to natural laws.

From Spencer, Hofstadter turns to a consideration of William Graham Sumner, the most influential American social Darwinist, whose writings glorified the competitive social order and justified existing social inequalities as the result of natural selection. Combining Darwinian ideas with the Protestant work ethic and classical economics, Sumner condemned any idea of government activism, preferring instead a complete "abnegation of state power." He offered defenders of the economic status quo a compelling rationale for opposing the demands of labor unions, Grangers, and others seeking to interfere with the "natural" workings of the social order.

Despite the book's title and the deftness with which he sketches the lineaments of social Darwinism in its opening chapters, Hofstadter actually devotes more attention to the theory's critics than its proponents. For a time, social Darwinism reigned supreme in American thought. But beginning in the 1880's, it came under attack from many sources—clergymen shocked by the inequities of the emerging industrial order and the harshness of unbridled competition, reformers proposing to unleash state activism in the service of social equality, and intellectuals of the emerging social sciences. Hofstadter makes no effort to disguise his distaste for the social Darwinists or his sympathy for the critics, especially the sociologists and philosophers who believed intellectuals could guide social progress (a view extremely congenial to Hofstadter at the time he was writing). In the 1880's, sociologist Lester Ward pointed out that economic competition bred not simply individual advancement but giant new corporations whose economic might needed to be held in check by government, and he ridiculed the social Darwinists' "fundamental error" that "the favors of the world are distributed entirely according to merit." But Hofstadter's true heroes were the early twentieth-century Pragmatists. William James destroyed Spencer's hold on philosophical thought by pointing to the elements of psychology—sentiment, emotion, and so on—ignored in the Darwinian model and by insisting that human intelligence enabled people to alter their own environment, thus rendering pointless the analogy with nature. James, however, evinced little interest in current events. Hofstadter identified more closely with John Dewey, whom he presents as a model of the socially responsible intellectual, the architect of a "new collectivism" in which an activist state attempts to guide and improve society.

By the turn of the century, social Darwinism was in full retreat. But even as Darwinian individualism waned, Darwinian ideas continued to influence social thinking in other ways. Rather than individuals striving for advancement, the struggling units of the analogy with nature became collectives—especially nations and races. With the United States emerging as a world power from the Spanish-American War, writers like John Fiske and Albert J. Beveridge marshaled Darwinian ideas in the service of imperialism, to legitimate the worldwide subordination of "inferior" races to Anglo-Saxon hegemony. In the eugenics movement that flourished in the early years of this century, Darwinism helped to underwrite the idea that immigration of less "fit" peoples was lowering the standard of American intelligence. Fortunately, the "racist-military" phase of social Darwinism was as thoroughly discredited by World War I, when it seemed uncomfortably akin to German militarism, as conservative individualism had been by the attacks of progressive social scientists.

When Hofstadter tries to *explain* the rise and fall of social Darwinism, he falls back on the base-superstructure model shared by Marxists and Beardians in the 1930's. Hofstadter recognizes that there was nothing inevitable in the appropriation of Darwinism for conservative purposes. Marx, after all, was so impressed by *The Origin of Species,* which dethroned revealed religion and vindicated the idea of progress through ceaseless struggle (struggle among classes, in his reading, rather than individuals), that he proposed to dedicate *Capital* to Darwin—an honor the latter declined. How then to account for the ascendancy, until the 1890's, of individualist, laissez-faire Darwinism? The reason, Hofstadter writes, was that social Darwinism served the needs of those groups that controlled the "raw, aggressive, industrial society" of the Gilded Age. Spencer, Sumner, and the other social Darwinians were telling businessmen and political leaders what they wanted to hear. Subsequently, it was not merely the penetrating criticism of Ward, Dewey, and others, but the middle class's growing disenchantment with unbridled competition, Hofstadter argues, that led it to repudiate social Darwinism and adopt a more reform-minded social outlook in the Progressive era.

Hofstadter's concluding thoughts amount to a reaffirmation both of the Beardian approach and of his own status as a radical intellectual. The rise and fall of social Darwinism, he writes, exemplified the "rule" that "changes in the structure of social ideas wait on general changes in economic and political life" and that ideas win wide acceptance based less on "truth and logic" than their "suitability to the intellectual needs and preconceptions of social interests." This, he adds, was "one of the great difficulties that must be faced by rational strategists of social change." Clearly, Hofstadter still viewed economic self-interest as the basis of political action, and clearly he identified with those "rational strategists of social change" who hoped to move the nation beyond social Darwinism's legacy.

Actually, Hofstadter offered no independent analysis of either the structure of American society or the ideas of most businessmen or politicians. His effort to explain social Darwinism's rise and fall is a kind of obiter dictum, largely confined to his brief concluding chapter. Indeed, Hofstadter later reflected that the book may have inadvertently encouraged the "intellectualist fallacy" by exaggerating the impact of ideas without placing them in the social context from which they sprang.[15] *Social Darwinism* is a work of intellectual history, not an examination of how ideas reflect economic structures. And as such, it retains much of its vitality half a century after it was written. The book's qualities would remain hallmarks of Hofstadter's subsequent writing—among them an amazing lucidity in presenting complex ideas, the ability to sprinkle his text with apt quotes that make precisely the right point, the capacity to bring past individuals to life in telling portraits. For a dissertation, it is a work of remarkable range, drawing not only on the writings of sociologists and philosophers but also on novels, treatises, sermons, and popular magazines to explore the debates unleashed by Darwinism. Very much a product of a specific moment in American history, it transcends the particulars of its origins to offer a compelling portrait of a critical period in the development of American thought. To the end of his life, Hofstadter's writings would center on *Social Darwinism's* underlying themes—

the evolution of social thought, the social context of ideologies, and the role of ideas in politics.

Social Darwinism has had an impact matched by few books of its generation. Hofstadter did not invent the term social Darwinism, which originated in Europe in the 1880's and crossed the Atlantic in the early twentieth century. But before he wrote, it was used only on rare occasions; he made it a standard shorthand for a complex of late-nineteenth-century ideas, a familiar part of the lexicon of social thought. The book demonstrates Hofstadter's ability, even in a dissertation, to move beyond the academic readership to address a broad general public. Since its appearance in a revised paperback edition in 1955 (Hofstadter left the argument unchanged but added an author's note and made several hundred "purely stylistic" alterations), it has sold more than 200,000 copies.[16]

Although, thanks to Hofstadter, social Darwinism has earned a permanent place in the vocabulary of intellectual history, his analysis has not escaped criticism. While few scholars have challenged Hofstadter's account of the main currents of late-nineteenth-century American thought, some have cast doubt on the extent of Darwin's influence on both laissez-faire conservatives and their liberal and radical critics. Soon after Hofstadter's revised edition appeared, Irvin G. Wyllie published an influential essay disputing Darwin's impact on American businessmen. Entrepreneurs, he found, justified the accumulation of wealth not by appealing to a vision of ruthless competition in which the success of some meant the ruin of others but by reference to hard work, Christian philanthropy, and the conviction that the creation of wealth benefited society as a whole.[17]

Since Hofstadter had devoted little attention to businessmen, apart from Andrew Carnegie, Wyllie's findings did not significantly affect the book's main argument. More damaging was the criticism advanced by Robert C. Bannister, who argued that Hofstadter had greatly exaggerated Darwin's influence on social thinkers themselves.[18] Remarkably few late-nineteenth-century writers, Bannister found, either invoked Darwin's authority, referred directly to biological evolution, or used Darwinian terminology such as survival of the fittest and the struggle for existence. The roots of their thought lay elsewhere, in classical economics and a preoccupation with defending property rights and limiting the power of the state. They were more likely to appeal to the authority of Adam Smith than Darwin, more likely to be influenced by contemporary events such as the 1877 railroad strike than by analogies to biological evolution. In fact, Bannister concluded, social Darwinism existed mainly as an "epithet," a label devised by advocates of a reforming state to stigmatize laissez-faire conservatism.

Hofstadter, to be sure, never claimed that Darwin created Gilded Age individualism; rather, he wrote, Darwinian categories supplemented an existing vocabulary derived from laissez-faire economics. Moreover, Bannister's definition of social Darwinism, requiring explicit use of Darwinian language, ignores less direct influences on social thought and more subtle adaptations of scientific reasoning. Toward the end of his life, Hofstadter praised his critic for careful reading of sources, but went on to suggest that "intellectual history, even as made by men who try to be rational and who try to regard distinctions, proceeds by more gross distinctions than you are aware of."[19] This was a fairly devastating critique of Bannister's approach (which, to his credit, Bannister included in the introduction to his own book). Nonetheless, Bannister's basic point struck home. Today, writers who examine Gilded Age conservatism are likely to locate its primary sources in realms other than Darwinism. Spencer's influence, it is true, still looms large; some have even suggested that the body of thought Hofstadter described ought to be called social Spencerism, not social Darwinism.[20]

This, however, would be a mistake, for if Hofstadter perhaps exaggerated Darwin's influence, he was certainly correct in identifying the idea that a science of society could be developed as all but ubiquitous among late-nineteenth- and early-twentieth-century intellectuals. Darwin's writings helped to catalyze this belief, which became a major point of self-definition and self-justification for intellectuals at a time when, through the rise of social science, their role in American society was becoming institutionalized. Hofstadter's central insight—that analogies with science helped to shape the way Americans perceived and interpreted issues from the differences between races and classes to the implications of state intervention in the economy—remains the starting point for serious investigations of American thought during the Gilded Age.[21]

Inevitably, *Social Darwinism* now seems in some ways dated. Today, in the wake of the "new social history," historians are more cognizant of the many groups that make up American society and no longer write confidently, as Hofstadter did, of a single "public mind." Given the pervasive impact of literary deconstruction, it seems decidedly (perhaps refreshingly) old-fashioned to assume, with Hofstadter, that texts have a single, rationally ascertainable meaning. But the most striking difference between Hofstadter's cast of mind and that of our own time lies in his resolute conviction that social Darwinism was an unfortunate but thankfully closed chapter in the history of social thought. Hofstadter wrote from the certainty that social Darwinism was demonstrably wrong, that biological analogies are "utterly useless" in understanding human society, that this episode had all been some kind of "ghastly mistake."

"A resurgence of social Darwinism . . . ," Hofstadter did note, was "always a possibility so long as there is a strong element of predacity in society." But he could hardly have foreseen the resurrection in the 1980's of biological explanations for human development[22] and of the social Darwinist mentality, if not the name itself: that government should not intervene to affect the "natural" workings of the economy, that the distribution of rewards within society reflects individual merit rather than historical circumstances, that the plight of the less fortunate, whether individuals or races, arises from their own failings. Had he lived to see social Darwinism's recrudescence, Hofstadter would certainly have noted how two previously distinct strands of this ideology have merged in today's conservatism—the laissez-faire individualism of a

William Graham Sumner (who, it should be noted, condemned the imperial state with the same vigor he applied to government intervention in the economy) and the militarist and racist Darwinism of the early twentieth century.

This is not the occasion for an extensive survey of Hofstadter's subsequent career,[23] but it may be useful to trace briefly how in later writings he both departed from and returned to the ideas expressed in his first book. If *Social Darwinism* announced Hofstadter as one of the most promising scholars of his generation, his second work, *The American Political Tradition*, published in 1948, propelled him to the very forefront of his profession. Since its appearance, this brilliant series of portraits of prominent Americans from the founding fathers through Jefferson, Jackson, Lincoln, and FDR has been a standard work in both college and high school history classes and been read by millions outside the academy. Hofstadter's insight was that virtually all his subjects held essentially the same underlying beliefs. Instead of persistent conflict (whether between agrarians and industrialists, capital and labor, or Democrats and Republicans), American history was characterized by broad agreement on fundamentals, particularly the virtues of individual liberty, private property, and capitalist enterprise. In *Social Darwinism*, he had observed that Spencer's doctrines came to America "long after individualism had become a national tradition." Now he appeared to be saying that the subject of his first book *was* the American political tradition.

With its emphasis on the ways an ideological consensus had shaped American development, *The American Political Tradition* in many ways marked Hofstadter's break with the Beardian and Marxist traditions. Along with Daniel Boorstin's *The Genius of American Politics* and Louis Hartz's *The Liberal Tradition in America* (both published a few years afterwards), Hofstadter's second book came to be seen as the foundation of the "consensus history" of the 1950's. But Hofstadter's writing never devolved into the uncritical celebration of the American experience that characterized much "consensus" writing. As Arthur Schlesinger, Jr., observed in a 1969 essay, there was a basic difference between *The American Political Tradition* and works like Boorstin's: "For Hofstadter [and, Schlesinger might have added, Hartz] perceived the consensus from a radical perspective, from the outside, and deplored it; while Boorstin perceived it from the inside and celebrated it." As a courtesy, Schlesinger sent Hofstadter a draft of the essay. In the margin opposite this sentence, Hofstadter, who never felt entirely comfortable with the consensus label, scribbled: "Thank you."[24]

Hofstadter had abandoned Beard's analysis of American development, but he retained his mentor's iconoclastic, debunking spirit. In Hofstadter's hands, Jefferson became a political chameleon, Jackson an exponent of liberal capitalism, Lincoln a mythmaker, and Roosevelt a pragmatic opportunist. And the domination of individualism and capitalism in American life produced not a benign freedom from "European" ideological conflicts, but a form of intellectual and political bankruptcy, an inability to think in original ways about the modern world. If the book has a hero, it is abolitionist Wendell Phillips, the only figure in *The American Political Tradition* never to hold political office. As in *Social Darwinism*, Hofstadter seemed to identify most of all with the engaged, reformist intellectual. It is indeed ironic that one of the most devastating indictments of American political culture ever written should have become the introduction to American history for two generations of students. One scholar at the time even sought to develop an alternative book of essays on America's greatest presidents precisely in order to counteract the "confusion and disillusionment" he feared Hofstadter was sowing among undergraduates.[25]

"All my books," Hofstadter wrote in the 1960's, "have been, in a certain sense, topical in their inspiration. That is to say, I have always begun with a concern with some present reality." His first two books, he went on, "refract the experiences of the depression era and the New Deal."[26] In the 1950's, a different "reality" shaped Hofstadter's writing—the Cold War and McCarthyism. In 1946, Hofstadter assumed a teaching position at Columbia and a few months later remarried, his first

wife having died in 1945. So he again found himself part of
New York's intellectual world. But this was very different
from the radical days of the 1930's. He had "grown a great
deal more conservative in the past few years," Hofstadter
wrote Merle Curti, then teaching at Wisconsin, in 1953.[27] Unlike many New York intellectuals, including a number of his
friends, Hofstadter never made a career of anticommunism.
Nor did he embrace neoconservatism, join the Congress for
Cultural Freedom, or become an apologist for America's Cold
War policies. He was repelled by McCarthyism (although he
declined Curti's invitation to condemn publicly the firing of
communist professors at the University of Washington).[28]
After supporting with "immense enthusiasm" Adlai Steven-
son's campaign for the White House in 1952, Hofstadter re-
treated altogether from politics. "I can no longer describe
myself as a radical, though I don't consider myself to be a conservative either," he wrote Harry Swados a decade after Ste-
venson's defeat. "I suppose the truth is, although my interests
are still very political, I none the less have no politics."[29]

What Hofstadter did have was a growing sense of the frag-
ility of intellectual freedom and social comity. His next book
was *The Development of Academic Freedom in the United States,*
written in collaboration with his Columbia colleague Walter P.
Metzger and published in 1955. As with other intellectuals,
his sensibility was strongly reinforced by the Holocaust in
Europe and the advent of McCarthyism at home. Hofstadter
understood McCarthysim not as a thrust for political advantage among conservatives seeking to undo the legacy of the
New Deal, but as the outgrowth of a deep-seated anti-intellectualism and provincialism in the American population. The
result was to deepen a distrust of mass politics that had been
simmering ever since he left the Communist party in 1939.
This attitude was reinforced by his search for alternative ways
of understanding political behavior. Reared on the assumption that politics essentially reflects economic interest, he now
became fascinated with alternative explanations of political
conduct—status anxieties, irrational hatreds, paranoia. Influenced by the popularity of Freudianism among New York

intellectuals of the 1950's and by his close friendships with the sociologist C. Wright Mills and literary critics Lionel Trilling and Alfred Kazin, Hofstadter became more and more sensitive to the importance of symbolic conduct, unconscious motivation, and, as he wrote in *The Age of Reform* (1955), the "complexities in our history which our conventional images of the past have not yet caught."

Hofstadter applied these insights to the history of American political culture in a remarkable series of books that made plain his growing conservatism and his sense of alienation from what he called America's periodic "fits of moral crusading." *The Age of Reform* offered an interpretation of Populism and Progressivism "from the perspective of our own time." In his master's essay, Hofstadter thoroughly sympathized with the struggles of the South's downtrodden tenant farmers. Now, he portrayed the Populists of the late nineteenth century as small entrepreneurs standing against the inevitable tide of economic development. He saw them as taking refuge in a nostalgic agrarian myth or lashing out against imagined enemies from British bankers to Jews in a precursor to "modern authoritarian movements." (Interestingly, this interpretation still bore the mark of the traditional Marxist critique of petty bourgeois social movements; the American Marxist thinker Daniel DeLeon had said much the same thing in the 1890's).

In *Social Darwinism*, William Graham Sumner and the capitalist plutocracy of the Gilded Age emerged as the main threats to American democracy; while noting the underside of Progressivism—its racism and Anglo-Saxonism—Hofstadter seemed to embrace its demand for state activism against social injustice. In *The Age of Reform*, he depicted the Progressives as a displaced bourgeoisie seeking in political reform a way to overcome their decline in status. Rather than a precursor of the New Deal, as it was commonly seen, Progressivism, with its infatuation with the idea of pure democracy, was the source of some of "our most troublesome contemporary delusions" about politics. A similar sensibility informed Hofstadter's next two books. In *Anti-Intellectualism in American Life*

(1963), he identified an American heartland "filled with people who are often fundamentalist in religion, nativist in prejudice, isolationist in foreign policy, and conservative in economics" as a persistent danger to intellectual life. In *The Paranoid Style in American Politics* (1965), he suggested that a common irrationality characterized popular enthusiasms of both the right and the left throughout American history.

The Age of Reform and *Anti-Intellectualism* won Hofstadter his two Pulitzer Prizes, but ironically today both seem more dated than his earlier books. Their deep distrust of mass politics, their apparent dismissal of the substantive basis of reform movements, strike the reader, even in today's conservative climate, as exaggerated and elitist. And since the rise of the new social history, it has become impossible to study mass movements without immersing oneself in local primary sources, rather than relying on the kind of imaginative reading of published works at which Hofstadter excelled. These books seemed to wed him to a consensus vision that deemed the American political system fundamentally sound and its critics essentially irrational.

Hofstadter's, however, was too protean an intellect to remain satisfied for long with the consensus framework. As social turmoil engulfed the country in the mid-1960's, Hofstadter remained as prolific as ever, but his underlying assumptions shifted again. In *The Progressive Historians* (1968), he attempted to come to terms once and for all with Beard and his generation. Their portrait of an America racked by perennial conflict, he noted, was overdrawn, but by the same token, the consensus outlook could hardly explain the American Revolution, Civil War, or other key periods of turmoil in the nation's past (including, by implication, the 1960's). *American Violence* (1970), a documentary volume edited with his graduate student Michael Wallace, offered a chilling record of political and social turbulence that utterly contradicted the consensus vision of a nation placidly evolving without serious disagreements. Finally, in *America at 1750*, the first volume in a proposed three-part narrative history of the nation, Hofstadter offered a portrait that brilliantly took account of the paradoxical coexistence of individual freedom and opportunity and widespread social injustice and human bondage in the colonial era. The book remained unfinished at the time of his death from leukemia in 1971, offering only a tantalizing suggestion of what his full account of the American past might have been.

It was during the 1960's that I became acquainted with Richard Hofstadter, first as my advisor for an undergraduate senior thesis and later as supervisor of my doctoral dissertation. There was a certain irony in our relationship. Today, I am fortunate enough to occupy the DeWitt Clinton chair in history at Columbia that Hofstadter once filled. Half a century ago, two years before I was born, the victim of political blacklisting Hofstadter replaced when he took his first full-time teaching position at City College was Jack D. Foner, my father.[30]

Whatever thoughts Hofstadter harbored about his particular twist of fate, he played brilliantly the role of intellectual mentor so critical to any student's graduate career. For all his accomplishments he was utterly without pretension, always unintimidating, never too busy to talk about one's work. Hofstadter's books directed me toward the subjects that have defined my own writing—the history of political ideologies and the interconnections between social development and political culture. Not that he tried to impose his own interests or views on his students—far from it. If no "Hofstadter school" emerged from Columbia, it is because he had no desire to create one. Indeed, it often seemed during the 1960's that his graduate students, many of whom were actively involved in the civil rights and antiwar movements, were having as much influence on his evolving interests and outlook as he on theirs. (The idea for the book on American violence, for example, originated with Michael Wallace.)

It would not be strictly true to call Hofstadter a great teacher. Writing was his passion, and he did not share the love of the classroom that marks the truly exceptional instructor. Hofstadter disliked the lecture podium intensely and almost seemed to go out of his way to make his lectures unappealing, perhaps to drive away some of the large numbers who inevitably registered for his courses. He was at his best in small seminars and individual consultations, and when criticizing written work. Here his erudition, open-mindedness, and desire to help each student do the best work of which he or she was capable came to the fore.

Despite his death at the relatively young age of fifty-four, Hofstadter left a prolific body of work, remarkable for its originality and readability, and his capacity to range over the length and breadth of American history. From *Social Darwinism* to *America at 1750*, his writings stand as a model of what historical scholarship at its finest can aspire to achieve.

ERIC FONER

NOTES

I wish to thank Jack D. Foner, Beatrice Kevitt Hofstadter, Walter P. Metzger, James P. Shenton, Fred Siegel, and Arthur W. Wang for their helpful comments and suggestions.

1. Richard Hofstadter, "The Great Depression and American History: A Personal Footnote," typescript of lecture, Box 36, Richard Hofstadter Papers, Rare Book and Manuscript Library, Columbia University. This lecture is undated, but internal evidence indicates that it was written in the mid-1960's.
2. Alfred Kazin, *New York Jew* (New York, 1978), 15. Susan S. Baker, *Radical Beginnings: Richard Hofstadter and the 1930s* (Westport, Conn., 1985) is the best account of Hofstadter's early career and his involvement with political radicalism.
3. Hofstadter to Kenneth Stampp, December 1944, quoted in Baker, *Radical Beginnings*, 180.
4. Hofstadter to Harvey Swados, January 20, April 30, 1938, Harvey Swados Papers, Archives, University of Massachusetts, Amherst.
5. *Ibid.*, May 29, 1938, February 16, 1939.
6. *Ibid.*, October 9, 1939, March 1940.
7. *Ibid.*, October 9, 1939.
8. Richard Hofstadter, "The Southeastern Cotton Tenants Under the AAA, 1933–1935" (Master's thesis, Columbia University, 1938).
9. David Hawke, "Interview: Richard Hofstadter," *History* 3 (1960), 141.
10. Richard Hofstadter, "The Tariff Issue on the Eve of the Civil War," *American Historical Review* 44 (October 1938), 50–55.
11. Felice Swados Hofstadter to Harvey Swados, February 6, 1939, Swados Papers.
12. Hofstadter to Harvey Swados, April 15, 1939, and Felice Swados Hofstadter to Harvey Swados, May 6, 1940, Swados Papers.
13. Hofstadter to Harvey Swados, May 1941, Swados Papers.

14. Felice Swados to Harvey Swados, June 2, 1940, with marginal comments by Richard Hofstadter, Swados Papers; Hofstadter, "The Great Depression and American History."
15. Richard Hofstadter, "Darwinism and Western Thought," in *Darwin, Marx, and Wagner*, ed. Henry L. Paine (Columbus, Ohio, 1962), 60–61.
16. Hawke, "Interview," 138. In the author's note, Hofstadter credits his second wife, Beatrice Kevitt Hofstadter, with "shar[ing] equally with me in the task of revision."
17. Irvin G. Wyllie, "Social Darwinism and the Businessman," *Proceedings of the American Philosophical Society* 103 (October 1959), 629–35.
18. Robert C. Bannister, *Social Darwinism: Science and Myth in Anglo-American Social Thought*, rev. ed. (Philadelphia, 1988).
19. Bannister, *Social Darwinism*, xviii.
20. Dorothy Ross, *The Origins of American Social Science* (New York, 1991), 85–91; Carl N. Degler, *In Search of Human Nature: The Fall and Revival of Darwinism in American Social Thought* (New York, 1991), 11.
21. Ross makes this point in *Origins of American Social Science*, as does John L. Recchuiti in "Intellectuals and Progressivism: New York's Social Scientific Community, 1880–1917" (Ph.D. diss., Columbia University, 1991). See also Nancy L. Stepan, "Race and Gender: The Role of Analogy in Science," *ISIS* 77 (1986), 261–77.
22. See Degler, *In Search of Human Nature*.
23. See Stanley Elkins and Eric McKitrick, "Richard Hofstadter: A Progress," in *The Hofstadter Aegis: A Memorial*, ed. Elkins and McKitrick (New York, 1974), 300–367, and Daniel J. Singal, "Beyond Consensus: Richard Hofstadter and American Historiography," *American Historical Review* 89 (October 1984), 976–1004.
24. Arthur M. Schlesinger, Jr., "Richard Hofstadter," in *Pastmasters*, ed. Robin W. Winks (New York, 1969) 289; Hofstadter's marginal comments on the draft of Schlesinger's essay are in the Hofstadter Papers.
25. Peter Novick, *That Noble Dream: The "Objectivity Question" and the American Historical Profession* (New York, 1988), 334n.
26. Hofstadter, "The Great Depression and American History."
27. Novick, *That Noble Dream*, 323.
28. Singal, "Beyond Consensus," 996n; Novick, *That Noble Dream*, 326.
29. Richard Hofstadter, *The Age of Reform* (New York, 1955) 14; Hofstadter to Harvey Swados, June 3, 1962, Swados Papers.
30. Hofstadter related this to me in 1962, when I first began working with him as an undergraduate. In 1981, New York City's Board of Higher Education apologized to the dismissed teachers, terming the events of 1941 an egregious violation of academic freedom.

Author's Note

This book was written during the years 1940–42 and first published in 1944. Although it was meant to be a reflective study rather than a tract for the times, it was naturally influenced by the political and moral controversy of the New Deal era. In some respects my views have changed since the first edition appeared. However, in revising the book I have not tried to bring the substance of the text into coherence with my present views, except at a few points where I was jarred by the rhetoric of the original version. After a period of years a book acquires an independent life, and the author may be so fortunate as to achieve a certain healthy detachment from it, which reconciles him to letting it stand on its own.

Although I have not made major changes of content, I have added a new introduction and completely rewritten the text, entering innumerable stylistic changes, rewriting several passages to rectify old errors or ambiguities and a few which, for reasons of taste, seemed to demand a complete reworking. Beatrice Kevitt Hofstadter shared equally with me in the task of revision.

The volume was originally completed while I held the William Bayard Cutting Traveling Fellowship at Columbia University and was first published under the auspices of the Albert J. Beveridge Memorial Fund of the American Historical Association.

R. H.

Social Darwinism
in American Thought

Author's Introduction

In 1959, one hundred years after the publication of Darwin's *The Origin of Species,* mankind has lived so long under the brilliant light of evolutionary science that we tend to take its insights for granted. It is hard for us fully to realize the immense thrill of enlightenment experienced by Darwin's own generation; it is harder still to appreciate the terrors experienced by the religiously orthodox among them. But John Fiske, the American evolutionist, put it well when he said that to have lived to see the old mists dissolve was "a rare privilege in the centuries."

Many scientific discoveries affect ways of living more profoundly than evolution did; but none have had a greater impact on ways of thinking and believing. In this respect, the space age does not promise even remotely to match it. Indeed, in all modern history there have been only a few scientific theories whose intellectual consequences have gone far beyond the internal development of science as a system of knowledge to revolutionize the fundamental patterns of thought. Discoveries of this magnitude shatter old beliefs and philosophies; they suggest (indeed often impose) the necessity of building new ones. They raise the promise—to some men infinitely alluring—of new and more complete systematizations of knowledge. They command so much interest and acquire so much prestige within the literate community that almost everyone feels obliged at the very least to bring his world-outlook into harmony with their findings, while some thinkers eagerly seize upon and enlist them in the formulation and propagation of their own views on subjects quite remote from science.

The first such episode in modern times, the formulation

of the Copernican system, required a major revision of cosmologies and opened up to learned men the fascinating and terrifying prospect that many long-received ideas about the world might have to be drastically revised. Once again, in the Newtonian and post-Newtonian eras, mechanical models of explanation began to be widely applied to the theory of man and to political philosophy, and the ideal of a science of man and of society took on new significance. Darwinism established a new approach to nature and gave fresh impetus to the conception of development; it impelled men to try to exploit its findings and methods for the understanding of society through schemes of evolutionary development and organic analogies. In our own time the work of Freud, whose insights originated and have their surest value in the sphere of clinical psychology and in the treatment of neuroses, has begun to be exploited in sociology and art, politics and religion.

Almost everywhere in western civilization, though in varying degrees according to intellectual traditions and personal temperaments, thinkers of the Darwinian era seized upon the new theory and attempted to sound its meaning for the several social disciplines. Anthropologists, sociologists, historians, political theorists, economists were set to pondering what, if anything, Darwinian concepts meant for their own disciplines. And if a great many intellectual gaucheries were committed in the course of this search for the consequences of Darwinism (as I believe there were), they were gaucheries to which we should be prepared to extend a certain measure of indulgence. The social-Darwinian generation, if we may call it that, was a generation that had to learn to live with and accommodate to startling revelations of possibly sweeping import; and neither the full meaning nor the limits of these revelations could be found until a great many thinkers had groped about, stumbled, and perhaps fallen in the dark.

The subject of this book is the effects of Darwin's work upon social thinking in America. In some respects the United States during the last three decades of the nineteenth and at the beginning of the twentieth century was *the* Darwinian country. England gave Darwin to the world, but the United States gave to Darwinism an unusually quick and sympathetic reception. Darwin was made an honorary member of the American Philosophical Society in 1869, ten years before his own university, Cambridge, awarded him an honorary degree. American scientists were prompt not only to accept the principle of natural selection but also to make important contributions to evolutionary science. The enlightened American reading public, which became fascinated with evolutionary speculation soon after the Civil War, gave a handsome reception to philosophies and political theories built in part upon Darwinism or associated with it. Herbert Spencer, who of all men made the most ambitious attempt to systematize the implications of evolution in fields other than biology itself, was far more popular in the United States than he was in his native country.

An age of rapid and striking economic change, the age during which Darwin's and Spencer's ideas were popularized in the United States was also one in which the prevailing political mood was conservative. Challenges to this dominant conservatism were never absent, but the characteristic feeling was that the country had seen enough agitation over political issues in the period before the Civil War, that the time had now come for acquiescence and acquisition, for the development and enjoyment of the great continent that was being settled and the immense new industries that were springing up.

Understandably Darwinism was seized upon as a welcome addition, perhaps the most powerful of all, to the store of ideas to which solid and conservative men appealed when they wished to reconcile their fellows to some of the hardships of life and to prevail upon them not to support hasty and ill-considered reforms. Darwinism was one of the great informing insights in this long phase in the history of the conservative mind in America. It was those who wished to defend the political status quo, above all the laissez-faire conservatives, who were first to pick up the instruments of social argument that were forged out of the Darwinian concepts. Only later, only after a style of social thought that can be called "social Darwinism" had taken clear and recognizable form, did the dissenters from this point of view move into the arena with formidable arguments. The most prominent dissenters, especially those like Lester Ward and the pragmatists who directed their criticism most immediately to the philosophical problems raised by social Darwinism, were thinkers who did not quarrel with the fundamental assumption that the new ideas had profound import for the theory of man and of society. They simply attempted to wrest Darwinism from the social Darwinists by showing that its psychological and social consequences could be read in totally different terms from those assumed by the more conservative thinkers who had preceded them in the field. Today their arguments seem, if not to everyone superior, at least to most of us an indispensable antidote to the plausible arguments of the men they criticized. But that they succeeded in establishing much of their critique should not lead us to forget that for many long years they represented a minority point of view. And hardly had they begun to succeed in showing that the individualist-competitive uses of Darwinism were open to question when a wholly new problem began to emerge, on which they were themselves unable to agree: a discussion arose over the question whether racist and imperialist invocations of Darwinism had any real justification.

Darwinism was used to buttress the conservative outlook in two ways. The most popular catchwords of Darwinism, "struggle for existence" and "survival of the fittest," when applied to the life of man in society, suggested that nature would provide that the best competitors in a competitive situation would win, and that this process would lead to continuing improvement. In itself this was not a new idea, as economists could have pointed out, but it did give the force of a natural law to the idea of competitive struggle. Secondly, the idea of development over aeons brought new force to another familiar idea in conservative political theory, the conception that all sound development must be slow and unhurried. Society could be envisaged as an organism (or as an entity something like an organism), which could change only at the glacial pace at which new species are produced in nature. One might, like William Graham Sumner, take a pessimistic view of the import of Darwinism, and conclude that Darwinism could serve only to cause men to face up to the inherent hardship of the battle of life; or one might, like Herbert Spencer, promise that, whatever the immediate hardships for a large portion of mankind, evolution meant progress and thus assured that the whole process of life was tending toward some very remote but altogether glorious consummation. But in either case the conclusions to which Darwinism was at first put were conservative conclusions. They suggested that all attempts to reform social processes were efforts to remedy the irremediable, that they interfered with the wisdom of nature, that they could lead only to degeneration.

As a phase in the history of conservative thought, social Darwinism deserves remark. In so far as it defended the status quo and gave strength to attacks on reformers and on almost all efforts at the conscious and directed change of society, social Darwinism was certainly one of the leading strains in American conservative thought for more than a generation. But it lacked many of the signal characteristics of conservatism as it is usually found. A conservatism that appealed more to the secularist than the pious mentality, it was a conservatism almost without religion. A body of belief whose chief conclusion was that the positive functions of the state should be kept to the barest minimum, it was almost anarchical, and it was devoid of that center of reverence and authority which the state provides in many conservative systems. Finally, and perhaps most important, it was a conservatism that tried to dispense with sentimental or emotional ties. Listen, for instance, to Willian Graham Sumner, explaining in his social-Darwinian classic, *What Social Classes Owe to Each Other*, what happened when men moved out of the medieval society based on status to the modern society based on contract:

> In the Middle Ages men were united by custom and prescription into associations, ranks, guilds, and communities of various kinds. These ties endured as long as life lasted. Consequently society was dependent, throughout all its details, on status, and the tie, or bond, was sentimental. In our modern state, and in the United States more than anywhere else, the social structure is based on contract, and status is of the least importance. Contract, however, is rational—even rationalistic. It is also realistic, cold, and matter-of-fact. A contract relation is based on a sufficient reason, not on custom or prescription. It is not permanent. It endures only so long as the reason for it endures. In a state based on contract sentiment is out of place in any public or common affairs. It is relegated to the sphere of private and personal relations. . . . The sentimentalists among us always seize upon the survivals of the old order. They want to save them and restore them. . . .
> Whether social philosophers think it desirable or not, it is out of the question to go back to status or to the sentimental relations which once united baron and retainer, master and servant, teacher and pupil, comrade and comrade. That we have lost some grace and elegance is undeniable. That life once held more poetry and romance is true enough. But it seems impossible that any one who has studied the matter should doubt that we have gained immeasurably, and that our further gains lie in going forward, not in going backward.

We may wonder whether, in the entire history of thought, there was ever a conservatism so utterly progressive as this. Some of the peculiarities of social Darwinism as a conservative rationale become apparent if one compares Sumner with Edmund Burke. As thinkers the two, of course, have something in common: both show the same resistance to attempts to break the mold of society and accelerate change; neither has any use for ardent reformers or revolutionaries, for the conception of natural rights, or for equalitarianism. But here the resemblance ends. Where Burke is religious, and relies upon an intuitive approach to politics and upon instinctive wisdom, Sumner is secularist and proudly rationalist. Where Burke relies upon the collective, long-range intelligence, the wisdom of the community, Sumner expects that individual self-assertion will be the only satisfactory expression of the wisdom of nature, and asks of the community

only that it give full play to this self-assertion. Where Burke
reveres custom and exalts continuity with the past, Sumner is
favorably impressed by the break made with the past when
contract supplanted status; he shows in this phase of his
work a disdain for the past that is distinctly the mark of a
culture whose greatest gift is a genius for technology. To
him it is only "sentimentalists" who want to save and restore
the survivals of the old order. Burke's conservatism seems
relatively timeless and placeless, while Sumner's seems to
belong pre-eminently to the post-Darwinian era and to
America.

Certainly in America the roles of the liberal and the conservative have been so often intermingled, and in some ways
reversed, that clear traditions have never taken form. This
will go far to reveal not only why our non-conservatives have
such a hard time explaining themselves today but also why
social Darwinism has such a peculiar ring as a conservative
social philosophy. In the American political tradition the
side of the "right"—that is, the side devoted to property
and less given to popular enthusiasms and democratic pro-
fessions—has been throughout the greater part of our history
identified with men who, while political conservatives, were
in economic and social terms headlong innovators and daring
promoters. From Alexander Hamilton through Nicholas Biddle to Carnegie, Rockefeller, Morgan, and their fellow ty-
coons, the men who held aristocratic or even plutocratic
views in matters political were also men who took the lead in
introducing new economic forms, new types of organization,
new techniques. If we look through the history of our practical politics for men who spoke favorably of restoring or
conserving old values we will find them—not exclusively,
to be sure, but most characteristically—among those who
leaned moderately to the "left." We find them among Jeffer-
sonians trying to save agrarianism and defend planter interests, among some Jacksonians pleading for a restoration of
republican simplicity, among Populists and Progressives trying to restore a popular democracy and a competitive economy
that they felt had formerly existed. The matter is, of course,

not entirely so simple as this, for the reformers espoused some techniques that were admittedly new in their efforts to achieve goals that were avowedly old. But it is not until the days of Franklin D. Roosevelt and the New Deal that the "liberal" or "progressive" side in American politics was also the side that was wholeheartedly identified with social and economic innovation and experiment—not until after almost 150 years of national development under the Constitution that the old pattern was completely broken.

I have said that social Darwinism was a secularist philosophy, but in one important respect this needs qualification. For social Darwinism of the hard-bitten sort represented by men like Sumner embodied a vision of life and, if the phrase will be admitted, expressed a kind of secular piety that commands our attention. Sumner, and no doubt after him all those who at one time or another were impressed by his views, were much concerned to face up to the hardness of life, to the impossibility of finding easy solutions for human ills, to the necessity of labor and self-denial and the inevitability of suffering. Theirs is a kind of naturalistic Calvinism in which man's relation to nature is as hard and demanding as man's relation to God under the Calvinistic system. This secular piety found its practical expression in an economic ethic that seemed to be demanded with special urgency by a growing industrial society which was calling up all the labor and capital it could muster to put to work on its vast unexploited resources. Hard work and hard saving seemed to be called for, while leisure and waste were doubly suspect. The economic ethic engendered by these circumstances put a premium on those qualities that seemed necessary for the disciplining of a labor force and a force of small investors. In articulating those needs, Sumner expressed an inherited conception of economic life, even today fairly widespread among conservatives in the United States, under which economic activity was considered to be above all a field for the development and encouragement of personal character. Economic life was construed as a set of arrangements that offered inducements to men of good character, while it punished those who were, in Sumner's words, "negligent, shiftless, inefficient, silly, and imprudent."

Today we have passed out of the economic framework in which that ethic was formed. We demand leisure; we demand that we be spared economic suffering; we build up an important business, advertising, whose function it is to encourage people to spend rather than save; we devise institutional arrangements like installment buying that permit people to spend what they have not yet earned; and we take up an economic theory like that of Keynes which stresses in a new way the economic importance of spending. We think of the economic order in terms of welfare and abundance rather than scarcity; we concern ourselves more with organization and efficiency than with character and punishments and rewards. One of the keys to the controversy of our time over the merits or defects of the "welfare state" is the fact that the very idea affronts the traditions of a great many men and women who were raised, if not upon the specific tenets of social Darwinism, at least upon the moral imperatives that it expressed. The growing divorcement of the economic process from considerations that can be used to discipline human character, and, still worse, our increasing philosophical and practical acceptance of that divorcement, is a source of real torment to the stern minority among us for whom the older economic ethic still has a great deal of meaning. And anyone who today imagines that he is altogether out of sympathy with that ethic should ask himself whether he has never, in contemplating the possibility of a nearly workless economic order, powered by atomic energy and managed by automation, had at least a moment of misgiving about the fate of man in a society bereft of the moral discipline of work.

It must also be conceded that, if men of Sumner's stamp seemed to contemplate human misery with callousness and with an excessively dogmatic certainty that nothing could be done about it, they tended to be stern masters of themselves where devotion to high principles was required. In this sense they had the virtue of consistency. Three times

Sumner himself put his position at Yale in jeopardy, by his uncompromising stand on the unpopular side of a controversy—once over the use of Spencer's work in teaching, once over his opposition to the protective tariff, and once over his denunciation of the Spanish-American War. And though the practical conclusions of their philosophy usually pleased the plutocrats, men of this stamp were not simple apologists of the plutocrats; nor can the values that meant most to them be described as plutocratic values. Sumner himself thought that the plutocrats were all too often greedy and irresponsible. The virtues that Spencer and Sumner preached—personal providence, family loyalty and family responsibility, hard work, careful management, and proud self-sufficiency—were middle-class virtues. There is a certain touching irony in the thought that, while writers like these preached slow change and urged men to adapt to the environment, the very millionaires whom they took to be the "fittest" in the struggle for existence were transforming the environment with incredible rapidity and rendering the values of the Spencers and Sumners of this world constantly less and less fit for survival.

RICHARD HOFSTADTER

Chapter One

The Coming of Darwinism

To have lived when this prodigious truth was advanced, debated, established, was a rare privilege in the centuries. The inspiration of seeing the old isolated mists dissolve and reveal the convergence of all branches of knowledge is something that can hardly be known to the men of a later generation, inheritors of what this age has won.

JOHN FISKE

When Charles Darwin's *The Origin of Species* dawned upon the world it aroused no such immediate furor in the United States as it did in England. A public sensation comparable to that stirred up in England by Huxley's famous clash with Wilberforce in June 1860 was impossible in America, where a critical election was beginning whose results would disrupt the Union and bring about a terrible Civil War. Although the first American edition of *The Origin of Species* was widely reviewed in 1860,[1] the coming of the war obscured new developments in scientific thought for all but professional scientists and a few hardy intellectuals.

Here and there, however, in quiet studies remote from the glare of politics, the ideas that were in time to transform the intellectual life of the country began to be cultivated. Darwin's friend Asa Gray, the Harvard botanist, after painstaking study of an advance copy of *The Origin of Species* sent to him by the author, wrote a careful review for the *American Journal of Sciences and Arts*, and with admirable foresight prepared a series of articles to defend evolution from the forthcoming charges of atheism. A few men who were already acquainted with the pre-Darwinian evolutionary speculation of Herbert Spencer were laying the foundations for a popular campaign in behalf of evolutionary science. A little-known resident of Salem named Edward

[1] Supernumerals refer to the notes at the end of this book (pages 217-242).

Silsbee, trying to arouse American interest in Spencer's ambitious project for a systematic philosophy, had found an immediate response in two men who would in time take the lead in remolding American thought. The first, John Fiske, a Harvard undergraduate who had already delved deeper than some of his professors into scientific and philosophical literature, went into ecstasies at the sight of Spencer's grandiose prospectus. The second, Edward Livingston Youmans, a popular lecturer on scientific subjects and author of a widely used textbook in chemistry, secured through his connection with D. Appleton and Company a sympathetic American publisher for Spencer's works.[2] When public attention turned to the problems raised by Darwinism, Fiske and Asa Gray led the movement to make evolution respectable, and Youmans became the self-appointed salesman of the scientific world-outlook.

Interest in the natural sciences grew rapidly. Articles in religious journals and popular magazines show that American readers were fast becoming absorbed in the evolution controversy during the years after the Civil War. To men of culture the idea of evolution, so startling to the popular mind, was hardly new. A man like Whitman, for example, could write of " this old theory, evolution, as broach'd anew, trebled, with indeed all devouring claims by Darwin." Some Americans were familiar with the historic tradition of speculative evolution, which had reached the point of violent controversy in the days of Cuvier, Geoffroy St. Hilaire, and Goethe.[3] Sir Charles Lyell's *Principles of Geology* (1832), which paved the way for the development hypothesis, had been widely read in the United States; and Robert Chambers' anonymously published *Vestiges of Creation* (American edition, 1845), a popular religious presentation of evolution, had received much attention.

The rise of biblical criticism and comparative religion, the general relaxation of fundamentalist faith encouraged by the liberal clergy, prepared many Americans for the acceptance of Darwinism. James Freeman Clarke's *Ten Great Religions*, a liberal study of world creeds, ran through

twenty-two editions in the fifteen years after its first appearance in 1871. A comparable popularization of the new
biblical scholarship appeared in 1891, when Washington
Gladden published *Who Wrote the Bible?* [4]

Many of the influences that brought independent thinkers to accept evolution were manifest in the early work of
John Fiske. Although Fiske came from a conventionally
religious New England family, his orthodoxy had been undermined by European science. Before entering Harvard
he had eagerly read Alexander von Humboldt's multi-vol-
umed *Cosmos,* an encyclopedic review of current attain-
ments of science, written in the language of naturalism. For
Fiske this book had been a revelation of almost religious in-
tensity, an emotional experience strong enough to force the
Civil War into the background. " What's a war," he wrote
in April 1861, " when a fellow has ' Kosmos ' on his shelf and
' Faust ' on his table? " [5] It was appropriate for Fiske to
couple Humboldt with Goethe. More than any other American of his time, Fiske felt the Faustian urge to devour the
entire realm of knowledge. This it was which sent him plow-
ing through the works of the English scientific writers, Mill,
Lewes, Buckle, Herschel, Bain, Lyell, and Huxley, impelled
him to undertake the most strenuous exercises in philology
(he had mastered eight languages and started six others by
the time he was twenty) , and kept him abreast of recent advances in biblical criticism. When Darwinism appeared,
with its imposing answer to the riddle of species, when Spencer promised a profound and authoritative interpretation of
the meaning of science, Fiske had long since changed gods.

Darwinism attracted many who lacked Fiske's ebullient
spirit and his freakish appetite for learning. To young
Henry Adams, bewildered by his recent experiences in Civil
War diplomacy, it at first suggested an intelligible rationale
for recent history:

He felt, like nine men in ten, an instinctive belief in Evolution . . .
Natural Selection led back to Natural Evolution, and at last to Natural
Uniformity. This was a vast stride. Unbroken Evolution under uni-
form conditions pleased everyone — except curates and bishops; it was

the very best substitute for religion; a safe, conservative, practical, thoroughly Common-Law deity. Such a working system for the universe suited a young man who had just helped to waste five or ten thousand million dollars and a million lives, more or less, to enforce unity and uniformity on people who objected to it; the idea was only too seductive in its perfection; it had the charm of art.[6]

For others, more confident of the optimistic implications of evolution, *The Origin of Species* became an oracle, consulted with the reverence usually reserved for Scripture. Charles Loring Brace, a leading social worker and reformer, read it thirteen times and emerged with the assurance that evolution guaranteed the final fruition of human virtue and the perfectibility of man. " For if the Darwinian theory be true, the law of natural selection applies to all the moral history of mankind, as well as the physical. Evil must die ultimately as the weaker element, in the struggle with good." [7]

Before it could secure a grip on the public mind and find a place in accepted patterns of thought, evolution had to prevail within the realm of science itself; and even scientists, especially those among the older generation who were committed to traditional ways of thinking, found adjustment to evolution a painful process. " It is like confessing murder," Darwin had remarked in 1844, when first he broached to Joseph Dalton Hooker his belief in the mutability of species. And Sir Charles Lyell, whose geology led to the very brink of the development hypothesis, hesitated for almost a decade before making the plunge.[8] Before Darwin, however, scientists had been puzzled by the inadequacy of the old notion of the fixity of species, which fitted so badly with the facts of paleontology and geology, with known fossil specimens, the wide variety of species, and the classification of living organisms. They had conventionally assumed that a series of acts of special creation had taken place. While this facile hypothesis may have accorded with their religious beliefs, the new generation of scientists, trained to see their function as a quest for natural causes, suspected that special creation was a poor intellectual makeshift. Among this generation the development hypothesis and the theory of natural selection spread rapidly, and a host of distinguished Darwinian advocates was soon in the field.

Alone among outstanding American naturalists, Louis Agassiz refused to the bitter end to accept Darwinism or evolution in any form.[9] Agassiz's master, Georges L. Cuvier, had been the leading opponent of evolution in the early nineteenth century, and the pupil fought Darwin as his teacher had fought Lamarck. To Agassiz, Darwinism was a crude and insolent challenge to the eternal verities, objectionable as science and abominable for its religious blasphemies. In his last article, published posthumously, Agassiz argued that all the evolution known to man is ontogenetic, the embryological development of the individual. Beyond this it would be impossible to go; evidence of descent of later from earlier species, or of the animal ancestry of man, was totally lacking. The classification of animals, Agassiz said, belied the idea of progression from lower to higher; the history of geological succession showed that the lowest in structure is not necessarily the first in time. A great diversity of animal types probably existed from the very beginning; it is therefore more likely that what men call species arose through separate successive acts of creation of differing individual organisms rather than natural selection or any other mode of purely natural development.[10]

Convinced that Darwinism was a fad (like Oken's *Naturphilosophie* in his younger days), Agassiz brashly asserted that he would " outlive this mania ";[11] but when he died in 1873 American science lost its last distinguished opponent of the new theory. Even if Agassiz had lived many years longer it is doubtful that his influence would have retarded the spread of evolution among scientists. Before his death his own students were falling away. Among them, Joseph Le Conte felt that outlines of the development theory were latent in Agassiz's own classification of animal forms, which need only be interpreted dynamically to yield a convincing picture of the evolutionary past.[12] William James, who had been intimate with Agassiz, was his bitterest critic. " The more I think of Darwin's ideas," he wrote to his brother

Henry in 1868, "the more weighty do they appear to me, though of course my opinion is worth very little — still, I *believe* that that scoundrel Agassiz is unworthy either intellectually or morally for him to wipe his shoes on, and I find a certain pleasure in yielding to the feeling." [13] Not long after Agassiz's death, a writer pointed out that eight of Agassiz's most eminent pupils at Harvard, including the master's own son, were evolutionists of relatively long standing.[14] In 1874, James Dwight Dana, dean of American geologists, published the final edition of his *Manual of Geology*, in which he too, after a prolonged attempt to resist natural selection, at last granted his endorsement.

Asa Gray soon found himself the acknowledged interpreter of American scientific opinion. Combining the conviction of a crusader with the caution of a scientist, Gray was peculiarly fitted to lead the Darwinian forces. His initial review of *The Origin of Species,* a brilliant essay upon the entire problem, had given American biologists a favorable but measured summary of Darwin's case. Conscientiously Gray had set forth what he conceived were the most cogent scientific objections to natural selection, but he praised the theory for its rigidly scientific contribution to biology. Darwin, he wrote guardedly, "has rendered a theory of derivation [of species] much less improbable than before. . . . Such a theory chimes in with the established doctrines of physical science, and is not unlikely to be largely accepted before it can be proved." With more daring he attacked Agassiz's theory of species as "theistic to excess" and praised Darwin's as an antidote. Closing on a note of defiance to possible religious criticisms, he declared that Darwinism is perfectly compatible with theism; and, he conceded, while it is also compatible with atheism, "that is true of physical theories generally." Natural selection, far from being an attack upon the argument from design in nature, may be considered one of the possible theories of the workings of God's plan.[15]

By the early 1870's the transmutation of species and natural selection dominated the outlook of American naturalists. At the twenty-fifth meeting of the American Association for the Advancement of Science, Vice-President Edward S. Morse gave a striking review of contributions of American biologists to evidences of evolution, which showed that their reception of Darwinism had been more than passive.[16] Most impressive among these were studies by Professor Othniel C. Marsh of Yale. An acquaintance of Gray, Lyell, and Darwin, one of the most colorful scientific men of the period, Marsh had set out in the early years of the decade in search of fossil specimens to confirm the development hypothesis. By 1874 he had collected a striking set of American fossil horses and published a paper, tracing the development of the horse through geologic ages, which Darwin later acclaimed as the best support of evolution appearing in the two decades following *The Origin of Species*.[17]

II

The conversion of the scientists promised early success in the universities, where the atmosphere was charged with electricity. A reform movement was under way to put greater stress upon science in the curricula, and science schools were being established to meet the country's growing need for technicians.[18] The appalling neglect of scientific specialization (which had brought forth in the smaller colleges such monstrosities as " Professor of Natural Philosophy, Chemistry, Mineralogy, and Geology and Lecturer of Zoölogy and Botany ") was now a patent anachronism in a nation that urgently needed science for its industry and agriculture and could well afford to patronize scientific development.

Harvard led the way in university reform with the appointment in 1869 of Charles William Eliot, a chemist, as its president. At Eliot's inauguration John Fiske privately expressed the hope that this appointment would signify the end of " old fogyism " at Harvard. The realization of this wish came sooner than he expected, and in a more personal way, when Eliot immediately called upon him to offer a series of special lectures at Harvard on the philosophy of science. Eight years before, as an undergraduate, Fiske had been threatened with dismissal from Harvard College if he were caught talking Comtism, which was generally considered atheistic. Now he was asked to hold forth at length under the aegis of the university on the positivist philosophy. Fiske, who had long since dropped Comte in favor of Spencer, undertook the task of defending Spencer against the charge of plagiarizing from Comte, but this hardly diminished the sweetness of his own vindication. The lectures, which were reported in the papers, aroused some criticism, but the audiences were large and enthusiastic.[19] Some years later, when William James used Spencer's *Principles of Psychology* as a textbook at Harvard, there was not a murmur of excitement. The new philosophy had made its way into the oldest of American universities quickly and almost without controversy.

At Yale it was again Spencer rather than Darwin who created the issue, which did not arise until 1879–80, when William Graham Sumner clashed with President Noah Porter. Porter, a Congregational clergyman, was not an uncompromising opponent of evolutionism in all its forms. Influenced by Professor Marsh's discoveries and his prestige, and impressed by the fine collection of specimens at Yale's own Peabody Museum, he had surrendered to evolution by 1877, when he gave an address in which he asserted that he found " no inconsistency between the findings of this museum on the one corner and the teachings of the college chapel on the other." [20] Nonetheless, he believed that American colleges should be kept " distinctively and earnestly Christian." When Sumner, also converted to evolution by Marsh's work, tried to use Spencer's *Study of Sociology* as a text in one of his courses, Porter objected to the work's antitheistic and anticlerical tone, and insisted that Sumner abandon it. A widely publicized controversy followed, which ended in a Pyrrhic victory for Porter.[21] Sumner, after excoriating Porter, threatened to resign, and was induced with some difficulty to remain. He dropped Spencer's book, on the ground that the controversy had undermined its value as a textbook, but otherwise continued in his independent ways. Porter himself conducted a course in " First Principles " to refute

Spencer's ideas, in which he used some of the evolutionist's writings. To his dismay, the appeal of Spencer's works was irresistible to many students, and they became converts to the doctrines Porter was at such pains to overthrow.[22]

Less prominent scholars and clerical teachers in some other schools of higher learning were neither as safe nor as successful as Fiske and Sumner. The geologist Alexander Winchell was dismissed from Vanderbilt in 1878, and occasional infringements of academic liberties in other institutions, North and South, caught the attention of the public throughout the 1880's and 1890's.[23] What is perhaps most noteworthy, however, is not the strength of the resistance but the rapidity with which the new ideas won their way in the better colleges and universities. Evolution penetrated the ranks of faculties and students alike. " Ten or fifteen years ago," declared Whitelaw Reid in an address at Dartmouth College in 1873, " the staple subject here for reading and talk, outside study hours, was English poetry and fiction. Now it is English science. Herbert Spencer, John Stuart Mill, Huxley, Darwin, Tyndall, have usurped the places of Tennyson and Browning, and Matthew Arnold and Dickens." [24]

The founding in 1876 of The Johns Hopkins University, an institution devoted to research and free of obligations to any religious denomination, marked a long forward step in higher education. Its first president, Daniel Coit Gilman, struck a symbolic note of defiance to obscurantism at its opening ceremonies by having Thomas Henry Huxley, who was in America on a lecture tour, give an address. Huxley's address was well received, but his appearance called forth the expected *odium theologicum*. " *It was bad enough to invite Huxley,*" wrote one divine. " *It were better to have asked God to be present. It would have been absurd to ask them both.*" [25] Such outcries, however, did not impede the development of the new institution, which was soon among the few leading universities in the advancement of scientific learning. Nor did the cries of alarm obscure or diminish Huxley's popularity; he found it necessary to refuse countless requests for lectures, and his comings and goings were reported with lavish care by the press.

Popular magazines promptly opened their columns to the evolution controversy. Characteristic progress over a ten-year period from hostility to skepticism to gingerly approval, and finally to full-blown praise, can be seen in the volumes of the *North American Review,* traditional forum of New England intellectuals. In 1860 an anonymous reviewer of *The Origin of Species,* arguing that natural selection would take an eternity of time to accomplish its task, rejected Darwin's theory as "fanciful." [26]   Four years later a writer pointed out that the development hypothesis, as a general conception, "has much to recommend it to the speculative mind. It is, as it were, an abstract statement of the order which the intellect expects to find in nature." [27]   In 1868, the freethinker Francis Ellingwood Abbot suggested that despite differences of opinion on minor points the development hypothesis would probably take a place among the accepted truths of science.[28]   In 1870 natural selection was praised by Charles Loring Brace as "one of the great intellectual events of the present century, influencing every department of investigation."   The following year the magazine published an essay by Chauncey Wright defending natural selection; this so impressed Darwin that he had it reprinted in pamphlet form for English readers.[29]

At the instance of Youmans, who saw the need for a popular magazine emphasizing scientific news, D. Appleton and Company founded *Appleton's Journal* in 1867. The journal was the first to run large numbers of articles on Spencer and Darwin and to provide regular publication for the popularizations of Youmans and Fiske. Neither wholly literary nor wholly scientific, *Appleton's Journal* pleased few readers.[30] More successful was the *Popular Science Monthly,* founded by Youmans in 1872. The monthly was surprisingly well received, considering the difficulty of some of its subject matter, and soon sold eleven thousand copies a month. There, next to more sensational sketches designed to satisfy common curiosity — "Great Fires and Rainstorms," "Hypnotism in Animals," "The Genesis of Superstition," "Earthquakes and Their Causes"—were learned articles on the philosophy of science, laudatory sketches of leading scientists, discussions of the reconciliation between science and religion, polemics against obscurantism, and reports on the latest progress of research. Edited on a high level and followed faithfully by a substantial body of readers, the monthly was the signal journalistic accomplishment of the scientific revival. To Youmans also must be given credit for organizing on behalf of Appleton's the famous International Scientific Series, a set of books by outstanding scientific figures of the time planned to cover almost the whole range of natural and social knowledge, which numbered among its contributors Walter Bagehot, John W. Draper, Stanley Jevons, Spencer, and Edward Tylor in the social sciences; Alexander Bain, Joseph Le Conte, Darwin, and Henry Maudsley in psychology and biology; and John Tyndall and others in physical science. Through the International Scientific Series, the *Popular Science Monthly,* and its control of American editions of Spencer's writings, Appleton dominated the new intellectual movement and rose to unchallenged leadership in the publishing world on the tidal wave of evolutionism.

The *Atlantic Monthly* also exploited the controversy by publishing Asa Gray's early defenses of Darwinism.[31] Seeking to maintain a noncommittal tone on Darwinism throughout the 1860's, the editors balanced the scale by printing one of Agassiz's counterblasts; but in 1872 an editorial on the rejection of Darwin by the French Academy of Science spoke of natural selection as having

. . . quite won the day in Germany and England, and very nearly won it in America. If the highest type of scientific mind be that which unites the power of originating grand generalizations with endless patience and caution in verifying them, then it is not too much to say that since the death of Newton this type has been in no one more perfectly realized than in Mr. Darwin.[32]

E. L. Godkin's *Nation* gave a favorable if none too conspicuous place to notices of evolutionary writings. Its reviewers were among the first to praise Darwin, Wallace, and Spencer. Gray's unsigned notices occasionally graced its columns, and some of his most vigorous onslaughts against recalcitrant naturalists and presumptuous clergymen appeared there. At a time when clerical magazines were in an uproar over Darwin's *The Descent of Man,* the *Nation* described it as "the most lucid and impartial exposition of the present state of scientific opinion respecting the origin of man and his relations to the lower animals." [33]

There could be no better testimony to an overwhelming interest in scientific developments and the new rationalism than the extensive daily newspaper coverage, in generous detail, of scientific or philosophical lectures. Fiske's Harvard lectures on "The Cosmic Philosophy" were reported in the New York *World* at the suggestion of the editor, Manton Marble. Huxley's lectures in New York were reprinted and discussed in the *Tribune,* and his visit was treated as ceremoniously as that of royalty.[34] It excited no surprise that George Ripley, one of the more vocal journalistic champions of Darwinism,[35] should take the dedication of the new *Tribune* building as an occasion for a muddy discussion of the metaphysical implications of nineteenth-century science.[36] The "universal drenching" of belles-lettres and journalism with natural selection amused an editor of the *Galaxy.* "Journalism is dyed so deep with it," he remarked, "that the favorite logic of the leading articles is 'survival of the fittest' and the favorite jest is 'sexual selection.'" He noticed that a Washington reporter for the *Herald* had recently done a sketch of the Senate in which members were portrayed in Darwinian terms as bulls, lions, foxes, and rats. At the latest New Orleans Mardi Gras the Missing Link had been used as a costume motif.[37]

III

The last citadels to be stormed were the churches, where evolution won its chief victories among the intellectually alert members of the more liberal Protestant denominations. Of course large numbers of devout persons, Protestant and Catholic, were untouched by Darwinism. Probably the most popular religious leader of the Gilded Age was the evangelist Dwight L. Moody, whose followers must have been blissfully ignorant about all of the troublesome questions raised by the new science. The persistence of fundamentalism into the twentieth century is a token of the incompleteness of the Darwinian conquest. Among the reflective congregations of the late nineteenth-century churches, however, there were vague emotional stirrings and intellectual dissatisfactions which helped to create a receptive frame of mind for a theology liberal enough to embrace the concept of evolution.[38]

Darwinism seemed to strike from more than one direction at the very heart of traditional theology. For nearly a century the argument from design, as popularized by the English theologian William Paley, had been standard proof of the existence of God. Now it seemed to many that Darwinism, by blasting at this theological foundation stone, must inevitably lead to atheism. The new theory also exploded traditional conceptions of sin, and with them the moral sanctions of the past. At the very least it clearly impaired the authority of Scripture by discrediting the Genesis version of the creation. Such was the initial orthodox reaction.[39] The appearance of *The Descent of Man* (1871) heaped fuel on the fires of clerical wrath,[40] for now human dignity itself was openly under attack. Religious readers pointed with horror at Darwin's too vivid description of man's ancestor as "a hairy quadruped, furnished with a tail and pointed ears, probably arboreal in habits."

Darwin's work and everything connected with it aroused virulent hostility throughout the 1860's and 1870's. Not a few of the clerical arguments were on the intellectual plane of the minister who asserted that Darwinism would be established only when scientists could take a monkey from the zoo and by natural selection make him into a man.[41] The tone became such that a clergyman, Professor W. N. Rice of Wesleyan University, remonstrated with his fellows against their attitude toward Darwin and suggested that they confine their criticism to the scientific issues.[42]

The most important clerical objection, of course, was that

Darwinism could not be reconciled with theism. Such was the central theme of the most popular exposition of anti-Darwinian views, Charles Hodge's *What Is Darwinism?* (1874). An old-school clergyman, author of one of the most imposing theological treatises of the time, and editor of the *Princeton Review,* Hodge could speak with authority for a large body of churchmen. In his polemical volume, Hodge reminded his readers that " the Bible has little charity for those who reject it. It pronounces them to be either derationalized or demoralized, or both." [43] The perilous paths of atheism threaten all who trifle with evolution, he declared, citing a formidable list of alleged materialists and atheists, including Darwin, Haeckel, Huxley, Büchner, and Vogt. With scant regard for facts,[44] Hodge charged that Darwin had carefully excluded any suggestion of design from nature, and closed with the assertion that Darwinism and atheism are synonymous.[45]

Catholic critics were often equally intransigent. Although mindful that St. George Mivart, an English Catholic and an able critic of natural selection, was an evolutionist, Orestes A. Brownson probably expressed the prevailing Catholic reaction when he urged a policy of no compromise with evolutionary biology. Dissatisfied with the weaker negations of both Protestant and many Catholic opponents of Darwinism, Brownson called for a categorical repudiation of nineteenth-century geology and biology, which he said represented a regression from the science of Aquinas. Lyell, Darwin, Huxley, Spencer, even Agassiz, came under his vigorous attack. " The *differentia* of man," he wrote, in an Aristotelian analysis of *The Descent of Man,* " not being in the ape, cannot be obtained from the ape by development. This sufficiently refutes Darwin's whole theory." The Genesis version of creation is still in possession, he concluded, and must be maintained until the contrary is fully demonstrated; the burden of proof therefore lies with Darwin.[46]

The most orthodox struggled with the desperation of men who felt their cause was doomed, but others retired to defensible positions in comparatively good order. The ultimate collapse of uncompromising opposition to evolution was foreshadowed as early as 1871, when James McCosh, the president of Princeton University and the semi-official voice of American Presbyterianism, acknowledged his acceptance of the development hypothesis in his *Christianity and Positivism*. An outstanding proponent of the current religious philosophy known as Scotch or " common sense " realism, and a man of unquestioned Christian integrity, McCosh had been specially imported from Scotland to give tone to Princeton. It was therefore a matter of considerable moment when, in a volume written to defend theism by the argument from design, he accepted the development hypothesis and conceded that natural selection is at least a portion of the truth:

> Darwinism cannot be regarded as settled. . . . I am inclined to think that the theory contains a large body of important truths which we see illustrated in every department of organic nature; but that it does not contain the whole truth, and that it overlooks more than it perceives. . . . That this principle [natural selection] is exhibited in nature and working to the advancement of plants and animals from age to age, I have no doubt. . . . But it has not been proven that there is no other principle at work.[47]

True, McCosh balked at the application of natural selection to mankind on the ground that a special act of creation explains more plausibly man's unique spiritual features; but the damage to orthodoxy was now done. Youmans wrote to Spencer in 1871:

> Things are going here furiously. I have never known anything like it. Ten thousand *Descent of Man* have been printed and I guess they are nearly all gone. . . . The progress of liberal thought is remarkable. Everybody is asking for explanations. The clergy are in a flutter. McCosh told them not to worry, as whatever might be discovered he would find design in it and put God behind it. Twenty-five clergymen of Brooklyn sent for me to meet them of a Saturday night and tell them what they should do to be saved. I told them they would find the way of life in the Biology and in the *Descent of Man*. They said " very good," and asked me to come again at the next meeting of the clerical club, to which I went and was again handsomely resoluted.[48]

The weekly *Independent*, the most influential religious paper in the country, with over six thousand clergymen on its mailing list, was among the first to give a relatively favorable hearing to evolution. Its initial review of *The Origin of Species* intimated that the book tended to displace the Creator from "the animated universe," but acknowledged the wealth of scientific material it contained. Subsequently the book was reccmmended for "the careful study of theologians and men of science." The paper was cautious and still under the influence of Agassiz in the late 1860's, although it had by then receded to the position that Darwinism would not affect theism, an acknowledgment which always served as an opening wedge. About this time, however, attempts at tenuous reconciliation of evolution with Scripture began to appear. "So long as the Bible does not assert that species were created distinct by an authoritative fiat we may be allowed to hear with no fluttering of our theologic nerves, the speculations of zoologists," wrote one reviewer.[49] By 1880 the *Independent* had completely reversed itself and had begun to publish full-throated polemics on behalf of evolution.[50] Other periodicals were slower to modify their views, but two decades after the introduction of Darwinism, some change was noticeable among even the most conservative.[51] The *New Englander*, an important forum of Yankee clergymen which had at first charged Darwin with reviving "an old, exploded theory," in 1883 published an interesting conciliatory article, in which the hysteria of some Christian apologists was admitted. "A fresh source of conviction," declared the writer, "is opened to our anticipations of immortality. It is the flattest inconsistency for an evolutionist to deny the probability of a higher future life."[52]

In the task of easing the transition to Darwinism for their brethren, liberal clergymen received aid and comfort from men of science. Asa Gray labored tirelessly to show that natural selection had no ultimate bearing on the argument from design, and that Darwin himself was explicitly theistic.[53] To those who insisted that the origins of species be left in the realm of the supernatural, Gray replied that they were arbitrarily limiting the field of science without enlarging that of religion. Joseph Le Conte, in his *Religion and Science*, a collection of Bible-class lectures, followed Gray in maintaining that the argument from design could not be changed by any possible answer to the question whether there had been transmutation of species or what the process of evolution might be. Science, he urged, should be looked upon not as the foe of religion, but rather as a complementary study of the ways in which the First Cause operated in the natural world. Whatever science might learn, the existence of God as First Cause could always be assumed.[54] Liberal theologians made good use of the fact that many of the advocates of evolution, like Le Conte, Dana, and McCosh, were men of undeniable Christian piety. They stood as personal symbols of the possibility of reconciling religion and science.[55]

The most important pulpit in the United States was brought within the evolutionary ranks when Henry Ward Beecher was converted, thanks to the combined impact of Darwin and Spencer. Through Beecher's *Christian Union,* which at one time reached a circulation of 100,000, and the *Outlook,* edited by Lyman Abbott, his successor at Plymouth Church, the liberalizing influence of Beecher's new theology was widespread. To the reconciliation of religion and science Beecher brought his national reputation, his brilliant, artful rhetoric, and the healthy good cheer of a man newly liberated from the confines of Puritan theology. His chief theoretical contribution was a carefully elaborated distinction between the science of theology and the art of religion: theology would be corrected, enlarged, and liberated by evolution, but religion, as a spiritual fixture in the character of man, would be unmoved.[56] Declaring himself "a cordial Christian evolutionist," Beecher publicly acknowledged Spencer as his intellectual foster father. It was Beecher who translated the solution of the design problem into the idioms of a business civilization, with the reminder that "design by wholesale is grander than design by retail."[57] Lyman Abbott agreed; moreover, he forswore the traditional notion of sin, which, he held, degraded God as well as man. He proposed to replace it with an evolutionary view in which every immoral act was to be regarded as a lapse into animality.

Sin would then be as abhorrent as ever, but the libel on God implied in the doctrine of original sin would be no more.[58]

By the 1880's, the lines of argument that would be taken in the reconciliation of science and religion had become clear. Religion had been forced to share its traditional authority with science, and American thought had been greatly secularized. Evolution had made its way into the churches themselves, and there remained not a single figure of outstanding proportions in Protestant theology who still ventured to dispute it. But evolution had been translated into divine purpose, and in the hands of skillful preachers religion was livened and refreshed by the infusion of an authoritative idea from the field of science. The ranks of the old foes soon could hardly be distinguished as they merged in common hostility to pessimism or skepticism about the promise of American life. The specter of atheism was no longer a menace, and surveys of the colleges where one would most expect to discover infidelity revealed how little there was.[59] With little exaggeration a minister could say that American infidelity had not produced " a single champion of cosmopolitan or even of national reputation." [60] " The spirit," explained Phillips Brooks, " that cries ' *Credo quia impossibile*,' the heroic spirit of faith, is too deep in human nature for any one century to eradicate it." [61] For was it not true, as Beecher told his Plymouth Church congregation, that " the moral structure of the human mind is such that it must have religion "? He continued:

It must have superstition, or it must have intelligent religion. It is just as necessary to men as reason is, as imagination is, as hope and desire are. Religious yearning is part and parcel of the human composition. And when you have taken down any theologic structure — if you should take down the Roman Church and scatter its materials; if then one by one you should dissect all Protestant theologies and scatter them — man would still be a religious animal, would need and be obliged to go about and construct some religious system for himself.[62]

To these sentiments of its leading divine, the Gilded Age gave unanimous consent.

Chapter Two

The Vogue of Spencer

As it seems to me, we have in Herbert Spencer not only the profoundest thinker of our time, but the most capacious and most powerful intellect of all time. Aristotle and his master were no more beyond the pygmies who preceded them than he is beyond Aristotle. Kant, Hegel, Fichte, and Schelling are gropers in the dark by the side of him. In all the history of science, there is but one name which can be compared to his, and that is Newton's . . .

F. A. P. BARNARD

I am an ultra and thoroughgoing American. I believe there is great work to be done here for civilization. What we want are ideas — large, organizing ideas — and I believe there is no other man whose thoughts are so valuable for our needs as yours are.

EDWARD LIVINGSTON YOUMANS to HERBERT SPENCER

"The peculiar condition of American society," wrote Henry Ward Beecher to Herbert Spencer in 1866, "has made your writings far more fruitful and quickening here than in Europe." [1] Why Americans were disposed to open their minds to Spencer, Beecher did not say; but there is much to substantiate his words. Spencer's philosophy was admirably suited to the American scene. It was scientific in derivation and comprehensive in scope. It had a reassuring theory of progress based upon biology and physics. It was large enough to be all things to all men, broad enough to satisfy agnostics like Robert Ingersoll and theists like Fiske and Beecher. It offered a comprehensive world-view, uniting under one generalization everything in nature from protozoa to politics. Satisfying the desire of "advanced thinkers" for a world-system to replace the shattered Mosaic cosmogony, it soon gave Spencer a public influence that transcended Darwin's. Moreover it was not a technical creed for professionals. Presented in language that tyros in philosophy could understand,[2] it made Spencer the metaphysician of the homemade intellectual, and the prophet of the cracker-barrel agnostic. Although its influence far outstripped its merits, the Spencerian system serves students of the American mind as a fossil specimen from which the intellectual body of the period may be reconstructed. Oliver Wendell Holmes hardly exaggerated when he expressed his doubt that " any writer of English except Darwin has done so much to affect our whole way of thinking about the universe." [3]

When Spencer's philosophy was winning its way in America, transcendentalism was in its twilight and the newer philosophical idealism inspired by Hegel was barely apparent on the horizon. Pragmatism was just emerging in the minds of Chauncey Wright and the little-appreciated Charles Peirce. The latter's now-famous article, " How to Make Our Ideas Clear," appeared in 1878, fourteen years after the first volume of Spencer's *Synthetic Philosophy;* and James's epoch-making California Union address, the opening gun in the campaign to popularize pragmatism, did not come until 1898. In the history of the American mind, however, the *Synthetic Philosophy* (which appeared in a series of volumes after 1860) is more than a colorless tenant of the vacancy between transcendentalism and pragmatism; although Emerson called Spencer a " stock writer " and James hurled at the Victorian Aristotle some of his sharpest barbs, Spencer was to most of his educated American contemporaries a great man, a grand intellect, a giant figure in the history of thought.

The ground for an American reception of Spencer's philosophy was well prepared in New England, which was, if one may judge by prominent persons among those answering Youmans' solicitations for advance subscribers to the volumes of the *Synthetic Philosophy,* the nursery of Spencerian influence. The presence on early subscription lists of such names as George Bancroft, Edward Everett, John Fiske, Asa Gray, Edward Everett Hale, James Russell Lowell, Wendell Phillips, Jared Sparks, Charles Sumner, and George Ticknor attests the power of New England intellectualism to provide for Spencer an American audience.[4] The effect of transcendentalism and Unitarianism in breaking up old orthodoxies and liberating the minds of American intellectuals cannot be measured but may certainly be sensed by any student of post-Civil War intellectual trends. Indeed, Americans were responsible for Spencer's chance to continue turning out the successive volumes of his project. In 1865, when the small returns from sales of his first volumes threatened to compel Spencer to give up his work, Youmans raised the necessary $7,000 among sympathetic Americans.[5]

Within a few years of his announcement of the *Synthetic Philosophy,* Spencer's work was known to a considerable body of American readers. The *Atlantic Monthly* commented in 1864:

Mr. Herbert Spencer is already a power in the world. . . . He has already influenced the silent life of a few thinking men whose belief marks the point to which the civilization of the age must struggle to rise. In America, we may even now confess our obligations to the writings of Mr. Spencer, for here sooner than elsewhere the mass feel as utility what a few recognize as truth. . . . Mr. Spencer represents the scientific spirit of the age. He makes note of all that comes within the range of sensuous experience, and declares whatever may be derived therefrom by careful induction. As a philosopher he does not go farther. . . . Mr. Spencer has already established principles which, however compelled for a time to compromise with prejudices and vested interests, will become the recognized basis of an improved society.[6]

In the three decades after the Civil War it was impossible to be active in any field of intellectual work without mastering Spencer.[7] Almost every American philosophical thinker of first or second rank — notably James, Royce, Dewey, Bowne, Harris, Howison, and McCosh — had to reckon with Spencer at some time. He had a vital influence upon most of the founders of American sociology, especially Ward, Cooley, Giddings, Small, and Sumner. " I imagine that nearly all of us who took up sociology between 1870, say, and 1890 did so at the instigation of Spencer," acknowledged Cooley. He continued:

His book, *The Study of Sociology,* perhaps the most readable of all his works, had a large sale and probably did more to arouse interest in the subject than any other publication before or since. Whatever we may have occasion to charge against him, let us set down at once a large credit for effective propagation.[8]

The Appleton publications, under the leadership of Youmans, pressed Spencer's interest incessantly, with the result that articles by him or about him were sprinkled throughout the popular magazines. The generation that acclaimed Grant as its hero took Spencer as its thinker. "Probably no other philosopher," wrote Henry Holt in later years,

> . . . ever had such a vogue as Spencer had from about 1870 to 1890. Most preceding philosophers had presumably been mainly restricted to readers habitually given to the study of philosophy, but not only was Spencer considerably read and generally talked about by the whole intelligent world in England and America, but that world was wider than any that preceded it.[9]

Spencer's impact upon the common man in the United States is impossible to gauge, although its effects are dimly perceptible. That he was widely read by persons who were partly or largely self-educated, by those who were laboriously plodding their way out of theological orthodoxy in a thousand towns and hamlets, is suggested by casual references to him in the lives of men who later achieved some fame. Theodore Dreiser, Jack London, Clarence Darrow, and Hamlin Garland have given intimations of Spencer's influence on their formative years. John R. Commons, in his autobiography, remarks on the fascination Spencer had for his father's friends during the writer's Indiana boyhood:

> He and his cronies talked politics and science. Every one of them in that Eastern section of Indiana was a Republican, living on the battle cries of the Civil War, and every one was a follower of Herbert Spencer, who was then the shining light of evolution and individualism. Several years later, in 1888, I was shocked, at a meeting of the American Economic Association, to hear Professor Ely denounce Herbert Spencer who had misled economists. I was brought up on Hoosierism, Republicanism, Presbyterianism, and Spencerism.[10]

The sales of Spencer's books in America from their earliest publication in the 1860's to December 1903 came to 368,755 volumes, a figure probably unparalleled for works in such difficult spheres as philosophy and sociology.[11] The number of persons who fell under his influence must be measured also by the extent to which copies were passed from hand to hand, and circulated through libraries. Of course it is impossible to say that the acceptance of his ideas was proportionate to their circulation. Certainly there was no lack of criticism. A *Nation* reviewer commented in 1884, before the vogue was over, that " the books examining or refuting Spencer now make an imposing library." [12] This criticism itself was another measure of the man's towering influence.

II

Herbert Spencer and his philosophy were products of English industrialism. It was appropriate that this spokesman of the new era should have trained to be a civil engineer, and that the scientific components of his thought — the conservation of energy and the idea of evolution — should have been indirectly derived from earlier observations in hydrotechnics and population theory. Spencer's was a system conceived in and dedicated to an age of steel and steam engines, competition, exploitation, and struggle.

Spencer was born in 1820 into a lower-middle-class, traditionally nonconformist English family; to these beginnings he ascribed his lifelong maniacal hatred of state power. In his early years on the staff of the *Economist*, a free-trade propaganda organ, he was associated for a short time with a Godwinian philosophical anarchist, Thomas Hodgskin, whose principles he apparently absorbed. Spencer's thinking took shape in the bright light of English science and positive thought, and his great *Synthetic Philosophy* was an amalgam of the nonconformism of his family and the scientific learning so prominent in his intellectual environment. Lyell's *Principles of Geology*, Lamarck's theory of development, Von Baer's law in embryology, Coleridge's conception of a universal pattern of evolution, the anarchism of Hodgskin, the laissez-faire principles of the Anti-Corn Law League, the gloomy prognoses of Malthus, and the conservation of energy were the elements from which Spencer composed his monolithic system. His social ideas are intelligible only in the setting of this philosophy; his social laws were but special cases of his general principles,[18] and much of the appeal of his social theories in America lay in their association with his synthetic integration of knowledge.

The aim of Spencer's synthesis was to join in one coherent structure the latest findings of physics and biology. While the idea of natural selection had been taking form in the mind of Darwin, the work of a series of investigators in thermodynamics had also yielded an illuminating generalization. Joule, Mayer, Helmholtz, Kelvin, and others had been exploring the relations between heat and energy, and had brought forth the principle of the conservation of energy which Helmholtz enunciated most clearly in his *Die Erhaltung der Kraft* (1847). The concept won general acceptance along with natural selection, and the convergence of the two discoveries upon the nineteenth-century mind was chiefly responsible for the enormous growth in the prestige of the natural sciences. Science, it was believed, had now drawn the last line in its picture of a self-contained universe, in which matter and energy were never destroyed but constantly changing form, whose varieties of organic life were integral, intelligible products of the universal economy. Previous philosophies paled into obsolescence much as pre-Newtonian philosophies had done in the eighteenth century. The transition to naturalism was marked by an efflorescence of mechanistic world-systems, whose trend is suggested by the names of Edward Büchner, Jacob Moleschott, Wilhelm Ostwald, Ernst Haeckel, and Herbert Spencer. Among these new thinkers, Spencer most resembled the eighteenth-century philosophers in his attempt to apply the implications of science to social thought and action.

The conservation of energy — which Spencer preferred to call " the persistence of force " — was the starting point of his deductive system. The persistence of force, manifested in the forms of matter and motion, is the stuff of human inquiry, the material with which philosophy must build. Everywhere in the universe man observes the incessant redistribution of matter and motion, rhythmically apportioned between evolution and dissolution. Evolution is the progressive integration of matter, accompanied by dissipation of motion; dissolution is the disorganization of matter accompanied by the absorption of motion. The life process is essentially evolutionary, embodying a continuous change from incoherent homogeneity, illustrated by the lowly protozoa, to coherent heterogeneity, manifested in man and the higher animals.[14]

From the persistence of force, Spencer inferred that anything which is homogeneous is inherently unstable, since the different effects of persistent force upon its various parts must cause differences to arise in their future development.[15] Thus the homogeneous will inevitably develop into the heterogeneous. Here is the key to universal evolution. This progress from homogeneity to heterogeneity — in the formation of the earth from a nebular mass, in the evolution of higher, complex species from lower and simpler ones, in the embryological development of the individual from a uniform mass of cells, in the growth of the human mind, and in the progress of human societies — is the principle at work in everything man can know.[16]

The final result of this process, in an animal organism or society, is the achievement of a state of equilibrium — a process Spencer called "equilibration." The ultimate attainment of equilibration is inevitable, because the evolutionary process cannot go on forever in the direction of increasing heterogeneity. "Evolution has an impassable limit." [17] Here the pattern of universal rhythm comes into play: dissolution follows evolution, disintegration follows integration. In an organism this phase is represented by death and decay, but in society by the establishment of a stable, harmonious, completely adapted state, in which "evolution can end only in the establishment of the greatest perfection and the most complete happiness." [18]

This imposing positivistic edifice might have been totally unacceptable in America, had it not also been bound up with an important concession to religion in the form of Spencer's doctrine of the Unknowable. The great question of the day was whether religion and science could be reconciled. Spencer gave not only the desired affirmative answer, but also an assurance for all future ages that, whatever science might learn about the world, the true sphere of religion — worship of the Unknowable — is by its very nature inviolable.[19]

To determined representatives of religious orthodoxy, Spencer's compromise was no more acceptable than that of Gray and Le Conte, and denunciations of his philosophy appeared frequently in the theological journals of the 1860's. Religious leaders who were willing to dally with liberalism, however, saw much in Spencer to praise. While thinkers like McCosh found the Unknowable too vague and uncomforting for faith and worship, some could identify it with God.[20] Still others found an analogy between his views of the transition from egoism to altruism and the preachings of Christian ethics.[21]

III

Spencer's supposition that a general law of evolution could be formulated led him to apply the biologic scheme of evolution to society. The principles of social structure and change, if the generalizations of his system were valid, must be the same as those of the universe at large. In applying evolution to society, Spencer, and after him the social Darwinists, were doing poetic justice to its origins. The " survival of the fittest " was a biological generalization of the cruel processes which reflective observers saw at work in early nineteenth-century society, and Darwinism was a derivative of political economy. The miserable social conditions of the early industrial revolution had provided the data for Malthus' *Essay on the Principle of Population,* and Malthus' observations had been the matrix of natural-selection theory. The stamp of its social origin was evident in Darwinian theory. " Over the whole of English Darwinism," Nietzsche once observed, " there hovers something of the odor of humble people in need and in straits." [22] Darwin acknowledged his great indebtedness to Malthus:

In October 1838, that is, fifteen months after I had begun my systematic inquiry, I happened to read for amusement " Malthus on Population," and being well prepared to appreciate the struggle for existence which everywhere goes on from long-continued observation of the habits of animals and plants, it at once struck me that under these circumstances favorable variations would tend to be preserved and unfavorable ones to be destroyed. The result of this would be the formation of new species.[23]

Alfred Russel Wallace, Darwin's co-discoverer of natural selection, likewise acknowledged that Malthus had given him " the long-sought clue to the effective agent in the evolution of organic species." [24]

Spencer's theory of social selection, also written under the stimulus of Malthus, arose out of his concern with population problems. In two famous articles that appeared in 1852, six years before Darwin and Wallace jointly published sketches of their theory, Spencer had set forth the view that the pressure of subsistence upon population must have a beneficent effect upon the human race. This pressure had been the immediate cause of progress from the earliest human times. By placing a premium upon skill, intelligence, self-control, and the power to adapt through technological innovation, it had stimulated human advancement and selected the best of each generation for survival.

Because he did not extend his generalization to the whole animal world, as Darwin did, Spencer failed to reap the full harvest of his insight, although he coined the expression " survival of the fittest." [25] He was more concerned with mental than physical evolution, and accepted Lamarck's theory that the inheritance of acquired characteristics is a means by which species can originate. This doctrine confirmed his evolutionary optimism. For if mental as well as physical characteristics could be inherited, the intellectual powers of the race would become cumulatively greater, and over several generations the ideal man would finally be developed. Spencer never discarded his Lamarckism, even when scientific opinion turned overwhelmingly against it.[26]

Spencer would have been the last to deny the primacy of ethical and political considerations in the formulation of his thought. " My ultimate purpose, lying behind all proximate purposes," he wrote in the preface to his *Data of Ethics,* " has been that of finding for the principles of right and wrong in conduct at large, a scientific basis." It is not surprising that he began his literary career with a book on ethics rather than metaphysics. His first work, *Social Statics* (1850), was an attempt to strengthen laissez faire with the imperatives of biology; it was intended as an attack upon Benthamism, especially the Benthamite stress upon the positive role of legislation in social reform. Although he consented to Jeremy Bentham's ultimate standard of value — the greatest happiness of the greatest number — Spencer discarded other phases of utilitarian ethics. He called for a return to natural rights, setting up as an ethical standard the right of every man to do as he pleases, subject only to the condition that he does not infringe upon the equal rights of others. In such a scheme, the sole function of the state is negative — to insure that such freedom is not curbed.

Fundamental to all ethical progress, Spencer believed, is the adaptation of human character to the conditions of life. The root of all evil is the " non-adaptation of constitution to conditions." Because the process of adaptation, founded in the very nature of the organism, is constantly at work, evil tends to disappear. While the moral constitution of the human race is still ridden with vestiges of man's original predatory life which demanded brutal self-assertion, adaptation assures that he will ultimately develop a new moral constitution fitted to the needs of civilized life. Human perfection is not only possible but inevitable:

The ultimate development of the ideal man is logically certain — as certain as any conclusion in which we place the most implicit faith; for instance that all men will die. . . . Progress, therefore, is not an accident, but a necessity. Instead of civilization being artificial, it is a part of nature; all of a piece with the development of the embryo or the unfolding of a flower.[27]

Despite its radicalism on incidental themes — the injustice of private land ownership, the rights of women and children, and a peculiar Spencerian " right to ignore the state " which was dropped from his later writings — the main trend of Spencer's book was ultra-conservative. His categorical repudiation of state interference with the "natural," unimpeded growth of society led him to oppose all state aid to the poor. They were unfit, he said, and should be eliminated. " The whole effort of nature is to get rid of such, to clear the world of them, and make room for better." Nature is as insistent upon fitness of mental character as she is upon physical character, "and radical defects are as much causes of death in the one case as in the other." He who loses his life because of his stupidity, vice, or idleness is in the same class as the victims of weak viscera or malformed limbs. Under nature's laws all alike are put on trial. " If they are sufficiently complete to live, they *do* live, and it is well they should live. If they are not sufficiently complete to live, they die, and it is best they should die." [28]

Spencer deplored not only poor laws, but also state-supported education, sanitary supervision other than the suppression of nuisances, regulation of housing conditions, and even state protection of the ignorant from medical quacks.[29] He likewise opposed tariffs, state banking, and government postal systems. Here was a categorical answer to Bentham.

In Spencer's later writings social selection was less prominent, although it never disappeared. The precise degree to which Spencer based his sociology upon biology was never a matter of common agreement, and the inconsistencies and ambiguities of his system gave rise to a host of Spencer exegesists, among whom the most tireless and sympathetic was Spencer himself.[30] Accused of brutality in his application of biological concepts to social principles, Spencer was compelled to insist over and over again that he was not opposed to voluntary private charity to the unfit, since it had an elevating effect on the character of the donors and hastened the development of altruism; he opposed only compulsory poor laws and other state measures.[31]

Spencer's social theory was more fully developed in the *Synthetic Philosophy*. In *The Principles of Sociology* there is a long exposition of the organic interpretation of society, in which Spencer traces the parallels between the growth, differentiation, and integration of society and of animal bodies.[32] Although the purposes of a social organism are different from those of an animal organism, he maintained that there is no difference in their laws of organization.[33] Among societies as among organisms, there is a struggle for existence. This struggle was once indispensable to social evolution, since it made possible successive consolidations of small groups into large ones and stimulated the earliest forms of social coöperation.[34] But Spencer, as a pacifist and internationalist, shrank from applying this analysis to contemporary society. In the future these intersocial struggles, he asserted, would lose their utility and die out. The very process of social consolidation brought about by struggles and conquest eliminates the necessity for continued conflict. Society then passes from its barbarous or militant phase into an industrial phase.

In the militant phase, society is organized chiefly for survival. It bristles with military weapons, trains its people for warfare, relies upon a despotic state, submerges the individual, and imposes a vast amount of compulsory coöperation. In contests among such societies those best exemplifying these militant traits will survive; and individuals best adapted to the militant community will be the dominating types.[35]

The creation of larger and larger social units through conquests by militant states widens the areas in which internal peace and application to the industrial arts become habitual. The militant type now reaches the evolutionary stage of equilibration. There emerges the industrial type of society, a regime of contract rather than status, which unlike the older form is pacific,[36] respectful of the individual, more heterogeneous and plastic, more inclined to abandon economic autonomy in favor of industrial coöperation with other states. Natural selection now works to produce a completely different individual character. Industrial society requires security for life, liberty, and property; the character type most consonant with this society is accordingly peaceful, independent, kindly, and honest. The emergence of a new human nature hastens the trend from egoism to altruism which will solve all ethical problems.[37]

Spencer emphasized that, in the interest of survival itself, coöperation in industrial society must be voluntary, not compulsory. State regulation of production and distribution, as proposed by socialists, is more akin to the organization of militant society, and would be fatal to the survival of the industrial community; it would penalize superior citizens and their offspring in favor of the inferior, and a society adopting such practices would be outstripped by others.[38]

In *The Study of Sociology*, first published in the United States in 1872–73 in serial form by the *Popular Science Monthly* and incorporated in the International Scientific Series, Spencer outlined his conception of the practical value of social science. Written to show the desirability of a naturalistic social science and to defend sociology from the criticisms of theologians and indeterminists, the book had a notable influence on the rise of sociology in the United States.[39] Spencer was animated by the desire to foster a science of society that would puncture the illusions of legislative reformers who, he believed, generally operated on the assumption that social causes and effects are simple and easily calculable, and that projects to relieve distress and remedy ills will always have the anticipated effect. A science of sociology, by teaching men to think of social causation scientifically, would awaken them to the enormous complexity of the social organism, and put an end to hasty legislative panaceas.[40] Fortified by the Darwinian conception of gradual modification over long stretches of time, Spencer ridiculed schemes for quick social transformation.

The great task of sociology, as Spencer envisioned it, is to chart "the normal course of social evolution," to show how it will be affected by any given policy, and to condemn all types of behavior that interfere with it.[41] Social science is a practical instrument in a negative sense. Its purpose is not to guide the conscious control of societal evolution, but rather to show that such control is an absolute impossibility, and that the best that organized knowledge can do is to teach men to submit more readily to the dynamic factors in progress. Spencer referred to the function of a true theory of society as a lubricant but not a motive power in progress: it can grease the wheels and prevent friction but cannot keep the engine moving.[42] " There cannot be more good done," he said, " than that of letting social progress go on unhindered; yet an immensity of mischief may be done in the way of disturbing, and distorting and repressing, by policies carried out in pursuit of erroneous conceptions." [43] Any adequate theory of society, Spencer concluded, will recognize the " general truths " of biology and will refrain from violating the selection principle by " the artificial preservation of those least able to take care of themselves." [44]

IV

With its rapid expansion, its exploitative methods, its desperate competition, and its peremptory rejection of failure, post-bellum America was like a vast human caricature of the Darwinian struggle for existence and survival of the fittest. Successful business entrepreneurs apparently accepted almost by instinct the Darwinian terminology which seemed to portray the conditions of their existence.[45] Businessmen are not commonly articulate social philosophers, but a rough reconstruction of their social outlook shows how congenial to their thinking were the plausible analogies of social selection, and how welcome was the expansive evolutionary optimism of the Spencerian system. In a nation permeated with the gospel of progress, the incentive of pecuniary success appealed even to many persons whose ethical horizons were considerably broader than those of business enterprise. " I perceive clearly," wrote Walt Whitman in *Democratic Vistas,* " that the extreme business energy, and this almost maniacal appetite for wealth prevalent in the United States, are parts of amelioration and progress, indispensably needed to prepare the very results I demand. My theory includes riches, and the getting of riches . . ." No doubt there were many to applaud the assertion of the railroad executive Chauncey Depew that the guests at the great dinners and public banquets of New York City represented the survival of the fittest of the thousands who came there in search of fame, fortune, or power, and that it was "superior ability, foresight, and adaptability" that brought them successfully through the fierce competitions of the metropolis.[46] James J. Hill, another railroad magnate, in an essay defending business consolidation, argued that "the fortunes of railroad companies are determined by the law of the survival of the fittest," and implied that the absorption of smaller by larger roads represents the industrial analogy to the victory of the strong.[47] And John D. Rockefeller, speaking from an intimate acquaintance with the methods of competition, declared in a Sunday-school address:

> The growth of a large business is merely a survival of the fittest. . . . The American Beauty rose can be produced in the splendor and fragrance which bring cheer to its beholder only by sacrificing the early buds which grow up around it. This is not an evil tendency in business. It is merely the working-out of a law of nature and a law of God.[48]

The most prominent of the disciples of Spencer was Andrew Carnegie, who sought out the philosopher, became his intimate friend, and showered him with favors. In his autobiography, Carnegie told how troubled and perplexed he had been over the collapse of Christian theology, until he took the trouble to read Darwin and Spencer.

> I remember that light came as in a flood and all was clear. Not only had I got rid of theology and the supernatural, but I had found the truth of evolution. "All is well since all grows better," became my motto, my true source of comfort. Man was not created with an instinct for his own degradation, but from the lower he had risen to the higher forms. Nor is there any conceivable end to his march to perfection. His face is turned to the light; he stands in the sun and looks upward.[49]

Perhaps it was comforting, too, to discover that social laws were founded in the immutable principles of the natural order. In an article in the *North American Review*, which he ranked among the best of his writings, Carnegie emphasized the biological foundations of the law of competition. However much we may object to the seeming harshness of this law, he wrote, "It is here; we cannot evade it; no substitutes for it have been found; and while the law may sometimes be hard for the individual, it is best for the race, because it insures the survival of the fittest in every department." Even if it might be desirable for civilization eventually to discard its individualistic foundation, such a change is not practicable in our age; it would belong to another " long succeeding sociological stratum," whereas our duty is with the here and now.[50]

The reception accorded to Spencer's social ideas cannot be dissociated from that accorded to the main body of his thought; however some part of his success probably came because he was telling the guardians of American society what they wanted to hear. Grangers, Greenbackers, Single Taxers, Knights of Labor, trade unionists, Populists, Socialists Utopian and Marxian — all presented challenges to the existing pattern of free enterprise, demanded reforms by state action, or insisted upon a thorough remodeling of the social order. Those who wished to continue in established ways were pressed for a theoretical answer to the rising voices of criticism. Said ironmaster Abram S. Hewitt:

> The problem presented to systems of religion and schemes of government is, to make men who are equal in liberty — that is, in political rights and therefore entitled to the ownership of property — content with that inequality in its distribution which must inevitably result from the application of the law of justice.[51]

This problem the Spencerian system could solve.

Conservatism and Spencer's philosophy walked hand in hand. The doctrine of selection and the biological apology for laissez faire, preached in Spencer's formal sociological writings and in a series of shorter essays, satisfied the desire of the select for a scientific rationale. Spencer's plea for absolute freedom of individual enterprise was a large philosophical statement of the constitutional ban upon interference with liberty and property without due process of law. Spencer was advancing within a cosmic framework the same general political philosophy which under the Supreme Court's exegesis of the Fourteenth Amendment served so brilliantly to turn back the tide of state reform. It was this convergence of Spencer's philosophy with the Court's interpretation of due process which finally inspired Mr. Justice Holmes (himself an admirer of Spencer) to protest that "the fourteenth Amendment does not enact Mr. Herbert Spencer's Social Statics." [52]

The social views of Spencer's popularizers were likewise conservative. Youmans took time from his promotion of science to attack the eight-hour strikers in 1872. Labor, he urged in characteristic Spencerian vein, must "accept the spirit of civilization, which is pacific, constructive, controlled by reason, and slowly ameliorating and progressive. Coercive and violent measures which aim at great and sudden advantages are sure to prove illusory." He suggested that, if people were taught the elements of political economy and social science in the course of their education, such mistakes might be avoided.[53] Youmans attacked the newly founded American Social Science Association for devoting itself to unscientific reform measures instead of a " strict and passionless study of society from a scientific point of view." Until the laws of social behavior are known, he declared, reform is blind; the Association might do better to recognize a sphere of natural, self-adjusting activity, with which government intervention usually wreaks havoc.[54] There was precious little scope for meliorist activities in the outlook of one who believed with Youmans that science shows " that we are born well, or born badly, and that whoever is ushered into existence at the bottom of the scale can never rise to the top because the weight of the universe is upon him." [55]

Acceptance of the Spencerian philosophy brought with it a paralysis of the will to reform. One day, some years after the publication of *Progress and Poverty,* Youmans in Henry George's presence denounced with great fervor the political corruption of New York and the selfishness of the rich in ignoring or promoting it when they found it profitable to do so. "What do you propose to do about it? " George asked. Youmans replied, " Nothing! You and I can do nothing at all. It's all a matter of evolution. We can only wait for evolution. Perhaps in four or five thousand years evolution may have carried men beyond this state of things." [56]

The peak of Spencer's American popularity probably was reached in the fall of 1882, when he made a memorable visit to the United States. In spite of his aversion to reporters, Spencer received much attention from the press, and hotel managers and railway agents competed for the privilege of serving him.[57] Finally yielding one synthetic "interview" with the gentlemen of the press, Spencer expressed (it was a slightly jarring note) his fear that the American character was not sufficiently developed to make the best use of its republican institutions.[58] The prospect for the future, however, was encouraging; from "biological truths," he told the reporters, he inferred that the eventual mixture of the allied varieties of the Aryan race forming the population would produce "a finer type of man than has hitherto existed." Whatever difficulties the Americans might have to surmount, they might "reasonably look forward to a time when they will have produced a civilization grander than any the world has known." [59]

The climax of the visit was a hastily arranged banquet at Delmonico's, which gave American notables an opportunity to pay personal tribute. The dinner was attended by leaders in American letters, science, politics, theology, and business. Spencer's message to this distinguished audience was somewhat disappointing. He had observed, he said, an excess of hurry and hard labor in the tempo of American life, too much of the gospel of work; his friends would ruin their constitutions with exertion. The guests rewarded this appeal against strenuosity with a strenuous round of fulsome tributes, which painfully embarrassed even the vain Spencer.[60] William Graham Sumner ascribed the foundations of sociological method to the guest of honor; Carl Schurz suggested that the Civil War might have been averted if the South had been familiar with his *Social Statics;* John Fiske asserted that his services to religion were as great as his services to science; and Henry Ward Beecher struck a rather incongruous note at the end of a hearty testimonial by promising to meet him once again beyond the grave.

However imperfect the appreciation of the guests for the niceties of Spencer's thought, the banquet showed how popular he had become in the United States. When Spencer was on the dock, waiting for the ship to carry him back to England, he seized the hands of Carnegie and Youmans. " Here," he cried to reporters, " are my two best American friends." [61] For Spencer it was a rare gesture of personal warmth; but more than this, it symbolized the harmony of the new science with the outlook of a business civilization.[62]

The rise of critical reformism in economics and sociology, of pragmatism in philosophy, and of other tendencies that undermined Spencer's vogue and displaced his ideas — this remains to be treated elsewhere. It is enough to say that, surviving until 1903, he outlived by many years the popularity of his works. In his old age he was aware that the current of the times was running against his preaching, and a visitor of this period reported finding him " grievously disappointed " at the neglect of his political doctrines, the decline of individualism, and the rise of socialist ideals.[63] " Herbert Spencer was a name to conjure with twenty-five years ago," taunted a religious observer in 1917. " But how the mighty are fallen! How little interest is shown in Herbert Spencer at the present time! " [64]

While it was true that for younger men Spencer's name no longer carried its old ring of authority, the writer had forgotten that men who were then in their maturity — the publicists, industrialists, teachers, and writers of the governing generation — had spent their youth with Spencer. Whatever had become of the Synthetic Philosophy, the mark of his evolutionary individualism was indelible. As late as 1915, the *Forum* had seen fit to reprint a collection of Spencer's individualistic essays, " The Man Versus the State," " The New Toryism," " The Coming Slavery," " Over-Legislation," " The Sins of Legislators," and others, along with commentaries by a galaxy of Republican Party luminaries brilliant enough to dispel all doubt of the vitality of Spencer's influence among outstanding national leaders.[65] Nicholas Murray Butler, Charles William Eliot, Representative Augustus P. Gardner, Elbert H. Gary, David Jayne Hill, Henry Cabot Lodge, Elihu Root, and Harlan Fiske Stone responded to the editor's request for contributions by " leaders of thought in America who know the tremendous value of Spencer's work in our social system." Hill's remark that he saw at work in this country the same fatal and illogical procedure that Spencer had been fighting in England, " namely, the gradual imposition of a new bondage in the name of freedom . . . the increasing subjection of the citizens to the growing tyranny of officialism," made it clear that the essays were being republished as a manifesto against Wilson's New Freedom.[66]

Spencer's doctrines were imported into the Republic long after individualism had become a national tradition. Yet in the expansive age of our industrial culture he became the spokesman of that tradition, and his contribution materially swelled the stream of individualism if it did not change its course. If Spencer's abiding impact on American thought seems impalpable to later generations, it is perhaps only because it has been so thoroughly absorbed.[67] His language has become a standard feature of the folklore of individualism. " You can't make the world all planned and soft," says the businessman of Middletown. " The strongest and best survive — that's the law of nature after all — always has been and always will be." [68]

Chapter Three

William Graham Sumner
Social Darwinist

Let it be understood that we cannot go outside of this alternative: liberty, inequality, survival of the fittest; not-liberty, equality, survival of the unfittest. The former carries society forward and favors all its best members; the latter carries society downwards and favors all its worst members.

WILLIAM GRAHAM SUMNER

The most vigorous and influential social Darwinist in America was William Graham Sumner of Yale. Sumner not only made a striking adaptation of evolution to conservative thought, but also effectively propagated his philosophy through widely read books and articles, and converted his strategic teaching post in New Haven into a kind of social-Darwinian pulpit. He provided his age with a synthesis which, though not quite so grand as Spencer's, was bolder in its stark and candid pessimism. Sumner's synthesis brought together three great traditions of western capitalist culture: the Protestant ethic, the doctrines of classical economics, and Darwinian natural selection. Correspondingly, in the development of American thought Sumner played three roles: he was a great Puritan preacher, an exponent of the classical pessimism of Ricardo and Malthus, and an assimilator and popularizer of evolution.[1] His sociology bridged the gap between the economic ethic set in motion by the Reformation and the thought of the nineteenth century, for it assumed that the industrious, temperate, and frugal man of the Protestant ideal was the equivalent of the " strong " or the " fittest " in the struggle for existence; and it supported the Ricardian principles of inevitability and laissez faire with a hard-bitten determinism that seemed to be at once Calvinistic and scientific.

Sumner was born in Paterson, New Jersey, on October 30, 1840. His father, Thomas Sumner, was a hard-working, self-educated English laborer who had come to America because his family's industry was disrupted by the growth of the factory system. He brought up his children to respect the traditional Protestant economic virtues, and his frugality left a deep impress upon his son William, who came in time to acclaim the savings-bank depositor as "a hero of civilization." [2] The sociologist later wrote of his father:

His principles and habits of life were the best possible. His knowledge was wide and his judgment excellent. He belonged to the class of men of whom Caleb Garth in *Middlemarch* is the type. In early life I accepted, from books and other people, some views and opinions which differed from his. At the present time, in regard to these matters, I hold with him and not with the others. [3]

The economic doctrines of the classical tradition which were current in his early years strengthened Sumner's paternal heritage. He came to think of pecuniary success as the inevitable product of diligence and thrift, and to see the lively capitalist society in which he lived as the fulfillment of the classical ideal of an automatically benevolent, free competitive order. At fourteen he had read Harriet Martineau's popular little volumes, *Illustrations of Political Economy,* whose purpose was to acquaint the multitude with the merits of laissez faire through a series of parables illustrating Ricardian principles. There he became acquainted with the wage-fund doctrine, and its corollaries: "Nothing can permanently affect the rate of wages which does not affect the proportion of population to capital"; and "combinations of laborers against capitalists . . . cannot secure a permanent rise of wages unless the supply of labour falls short of the demand — in which case, strikes are usually unnecessary." There also he found fictional proof that "a self-balancing power being . . . inherent in the entire system of commercial exchange, all apprehensions about the results of its unimpeded operations are absurd," and that "a sin is committed when Capital is diverted from its normal course to be employed in producing at home that which is expen-

sive and inferior, instead of preparing that which will pur-
chase the same article cheaper and superior abroad." Charities, whether public or private, Miss Martineau held, would
never reduce the number of the indigent, but would only encourage improvidence and nourish " peculation, tyranny,
and fraud." ⁴ Later Sumner declared that his conceptions
of " capital, labor, money and trade were all formed by those
books which I read in my boyhood." ⁵ Francis Wayland's
standard text in political economy, which he recited in college, seems to have impressed him but little, perhaps because
it only confirmed well-fixed beliefs.

In 1859, when he matriculated at Yale, young Sumner devoted himself to theology. During undergraduate years Yale
was still a pillar of orthodoxy, dominated by its versatile president, Theodore Dwight Woolsey, who had just turned from
classical scholarship to write his *Introduction to the Study of
International Law*, and by the Rev. Noah Porter, Professor
of Moral Philosophy and Metaphysics, who as Woolsey's successor would one day cross swords with Sumner over the
proper place of the new science in education. Sumner, a
somewhat frigid youth (who could seriously ask, " Is the
reading of fiction justifiable? ") repelled many of his school-
mates; but his friends made up in munificence what they
lacked in numbers. One of them, William C. Whitney,
persuaded his elder brother Henry to supply funds for Sumner's further education abroad; and the Whitneys secured a
substitute to fill his place in the Union Army while Sumner
pursued theological studies at Geneva, Göttingen, and Ox-
ford.⁶ In 1868 Sumner was elected to a tutorship at Yale,
beginning a lifelong association with its faculty that would be
broken only by a few years spent as editor of a religious newspaper and rector of the Episcopal Church in Morristown,
New Jersey. In 1872 he was elevated to the post of Professor
of Political and Social Science in Yale College.

Despite personal coldness and a crisp, dogmatic classroom
manner, Sumner had a wider following than any other
teacher in Yale's history.⁷ Upperclassmen found unique satisfaction in his courses; lowerclassmen looked forward to

promotion chiefly as a means of becoming eligible to enroll in them.[8]   William Lyon Phelps, who took every one of Sumner's courses as a matter of principle without regard for his interest in the subject matter, has left a memorable picture of Sumner's dealings with a student dissenter:

> "Professor, don't you believe in any government aid to industries?"
> "No! it's root, hog, or die."
> "Yes, but hasn't the hog got a right to root?"
> "There are no rights. The world owes nobody a living."
> "You believe then, Professor, in only one system, the contract-competitive system?"
> "That's the only sound economic system. All others are fallacies."
> "Well, suppose some professor of political economy came along and took your job away from you. Wouldn't you be sore?"
> "Any other professor is welcome to try. If he gets my job, it is my fault. My business is to teach the subject so well that no one can take the job away from me."[9]

The stamp of his early religious upbringing and interests marked all Sumner's writings. Although clerical phraseology soon disappeared from his style, his temper remained that of a proselytizer, a moralist, an espouser of causes with little interest in distinguishing between error and iniquity in his opponents. "The type of mind which he exhibited," writes his biographer, "was the Hebraic rather than the Greek. He was intuitive, rugged, emphatic, fervently and relentlessly ethical, denunciatory, prophetic."[10]   He might insist that political economy was a descriptive science divorced from ethics,[11] but his strictures on protectionists and socialists resounded with moral overtones. His popular articles read like sermons.

Sumner's life was not entirely given to crusading. His intellectual activity passed through two overlapping phases, marked by a change less in his thought than in the direction of his work. During the 1870's, 1880's and early 1890's, in the columns of popular journals and from the lecture platform, he waged a holy war against reformism, protectionism, socialism, and government interventionism. In this period he published *What Social Classes Owe to Each Other* (1883), "The Forgotten Man" (1883), and "The Absurd

Effort to Make the World Over" (1894). In the early 1890's, however, Sumner turned his attention more and more to academic sociology. It was during this period that the manuscript of " Earth Hunger " was written and the monumental *Science of Society* projected. When Sumner, always a prodigious worker, found that his chapter on human customs had grown to 200,000 words, he decided to publish it as a separate volume. Thus, almost as an afterthought, *Folkways* was brought out in 1906.[12] Although the deep ethical feelings of Sumner's youth gave way to the sophisticated moral relativism of his social-science period, his underlying philosophy remained the same.

II

The major premises of his social philosophy Sumner derived from Herbert Spencer. For years after his graduate residence at Oxford, Sumner had had " vague notions floating in my head " about the possibility of creating a systematic science of society. In 1872, when *The Study of Sociology* was running serially in the *Contemporary Review*, Sumner seized upon Spencer's ideas, and the evolutionary viewpoint in social science captivated his mind. It seemed that Spencer's proposals showed the full potentialities of his own germinal ideas. The young man who had been impervious to Spencer's *Social Statics* (because " I did not believe in natural rights or in his ' fundamental principles ' ") now found *The Study of Sociology* irresistible. " It solved the old difficulty about the relations of social science to history, rescued social science from the dominion of cranks, and offered a definite and magnificent field to work, from which we might hope at last to derive definite results for the solution of social problems." After a few years, Professor O. C. Marsh's researches in the evolution of the horse fully convinced Sumner of the development hypothesis. Plunging into Darwin, Haeckel, Huxley, and Spencer, he saturated himself with evolutionism.[13]

Like Darwin before him, Sumner went to Malthus for the first principles of his system. In many respects his sociology simply retraced the several steps in biological and social reasoning which ran from Malthus to Darwin and through Spencer to the modern social Darwinists. The foundation of human society, said Sumner, is the man-land ratio. Ultimately men draw their living from the soil, and the kind of existence they achieve, their mode of getting it, and their mutual relations in the process are all determined by the proportion of population to the available soil.[14] Where men are few and soil is abundant, the struggle for existence is less savage, and democratic institutions are likely to prevail. When population presses upon the land supply, earth hunger arises, races of men move across the face of the world, militarism and imperialism flourish, conflict rages — and in government aristocracy dominates.

As men struggle to adjust themselves to the land, they enter into rivalry for leadership in the conquest of nature. In Sumner's popular essays he stressed the idea that the hardships of life are incidents of the struggle against nature, that " we cannot blame our fellow-men for our share of these. My neighbor and I are both struggling to free ourselves from these ills. The fact that my neighbor has succeeded in this struggle better than I constitutes no grievance for me." [15] He continued:

Undoubtedly the man who possesses capital has a great advantage over the man who has no capital at all in the struggle for existence. . . . This does not mean that one man has an advantage *against* the other, but that, when they are rivals in the effort to get the means of subsistence from Nature, the one who has capital has immeasurable advantages over the other. If it were not so capital would not be formed. Capital is only formed by self-denial, and if the possession of it did not secure advantages and superiorities of a high order men would never submit to what is necessary to get it.[16]

Thus the struggle is like a whippet race; the fact that one hound chases the mechanical hare of pecuniary success does not prevent the others from doing the same.

Sumner was perhaps inspired to minimize the human conflicts in the struggle for existence by a desire to dull the resentment of the poor toward the rich. He did not at all

times, however, shrink from a direct analogy between animal struggle and human competition.[17] In the Spencerian
intellectual atmosphere of the 1870's and 1880's it was natural for conservatives to see the economic contest in competitive society as a reflection of the struggle in the animal
world. It was easy to argue by analogy from natural selection of fitter organisms to social selection of fitter men, from
organic forms with superior adaptability to citizens with a
greater store of economic virtues. The competitive order was
now supplied with a cosmic rationale. Competition was glorious. Just as survival was the result of strength, success was
the reward of virtue. Sumner had no patience with those who
would lavish compensations upon the virtueless. Many economists, he declared (in a lecture given in 1879 on the effect
of hard times on economic thinking),

. . . seem to be terrified that distress and misery still remain on earth
and promise to remain as long as the vices of human nature remain.
Many of them are frightened at liberty, especially under the form of
competition, which they elevate into a bugbear. They think it bears
harshly on the weak. They do not perceive that here "the strong"
and "the weak" are terms which admit of no definition unless they
are made equivalent to the industrious and the idle, the frugal and the
extravagant. They do not perceive, furthermore, that if we do not
like the survival of the fittest, we have only one possible alternative,
and that is the survival of the unfittest. The former is the law of civilization; the latter is the law of anti-civilization. We have our choice
between the two, or we can go on, as in the past, vacillating between
the two, but a third plan — the socialist desideratum — a plan for nour-
ishing the unfittest and yet advancing in civilization, no man will ever
find.[18]

The progress of civilization, according to Sumner, depends
upon the selection process; and that in turn depends upon
the workings of unrestricted competition. Competition is
a law of nature which "can no more be done away with than
gravitation,"[19] and which men can ignore only to their
sorrow.

III

The fundamentals of Sumner's philosophy had been set
forth in his magazine articles long before his sociological

works were written. The first fact in life, he asserted, is the
struggle for existence; the greatest forward step in this struggle is the production of capital, which increases the fruitful-
ness of labor and provides the necessary means of an advance
in civilization. Primitive man, who long ago withdrew from
the competitive struggle and ceased to accumulate capital
goods, must pay with a backward and unenlightened way of
life.[20]   Social advance depends primarily upon hereditary
wealth; for wealth offers a premium to effort, and hereditary
wealth assures the enterprising and industrious man that he
may preserve in his children the virtues which have enabled
him to enrich the community. Any assault upon hereditary
wealth must begin with an attack upon the family and end
by reducing men to " swine." [21]   The operation of social selection depends upon keeping the family intact. Physical
inheritance is a vital part of the Darwinian theory; the
social equivalent of physical inheritance is the instruction of
the children in the necessary economic virtues.[22]

If the fittest are to be allowed to survive, if the benefits of
efficient management are to be available to society, the cap-
tains of industry must be paid for their unique organizing
talent.[23]   Their huge fortunes are the legitimate wages of
superintendence; in the struggle for existence, money is the
token of success. It measures the amount of efficient management that has come into the world and the waste that has
been eliminated.[24]   Millionaires are the bloom of a competitive civilization:

> The millionaires are a product of natural selection, acting on the
> whole body of men to pick out those who can meet the requirement
> of certain work to be done. . . . It is because they are thus selected
> that wealth — both their own and that entrusted to them — aggregates
> under their hands. . . . They may fairly be regarded as the naturally
> selected agents of society for certain work. They get high wages and
> live in luxury, but the bargain is a good one for society. There is the
> intensest competition for their place and occupation. This assures us
> that all who are competent for this function will be employed in it, so
> that the cost of it will be reduced to the lowest terms.[25]

In the Darwinian pattern of evolution, animals are unequal; this makes possible the appearance of forms with finer

adjustment to the environment, and the transmission of such superiority to succeeding generations brings about progress. Without inequality the law of survival of the fittest could have no meaning. Accordingly, in Sumner's evolutionary sociology, inequality of powers was at a premium.[26] The competitive process " develops all powers that exist according to their measure and degree." If liberty prevails, so that all may exert themselves freely in the struggle, the results will certainly not be everywhere alike; those of " courage, enterprise, good training, intelligence, perseverance " will come out at the top.[27]

Sumner concluded that these principles of social evolution negated the traditional American ideology of equality and natural rights. In the evolutionary perspective, equality was ridiculous; and no one knew so well as those who went to school to nature that there are no natural rights in the jungle. " There can be no rights against Nature except to get out of her whatever we can, which is only the fact of the struggle for existence stated over again." [28] In the cold light of evolutionary realism, the eighteenth-century idea that men were equal in a state of nature was the opposite of the truth; masses of men starting under conditions of equality could never be anything but hopeless savages.[29] To Sumner rights were simply evolving folkways crystallized in laws. Far from being absolute or antecedent to a specific culture — an illusion of philosophers, reformers, agitators, and anarchists — they are properly understood as " rules of the game of social competition which are current now and here." [30] In other times and places other mores have prevailed, and still others will emerge in the future:

Each set of views colors the *mores* of a period. The eighteenth-century notions about equality, natural rights, classes and the like produced nineteenth-century states and legislation, all strongly humanitarian in faith and temper; at the present time the eighteenth-century notions are disappearing, and the *mores* of the twentieth century will not be tinged by humanitarianism as those of the last hundred years have been.[31]

Sumner's resistance to the catchwords of the American tradition is also evident in his skepticism about democracy.

The democratic ideal, so alive in the minds of men as diverse as Eugene Debs and Andrew Carnegie, as a thing of great hopes, warm sentiments, and vast friendly illusions, was to him simply a transient stage in social evolution, determined by a favorable quotient in the man-land ratio and the political necessities of the capitalist class.[32]   " Democracy itself, the pet superstition of the age, is only a phase of the all-compelling movement. If you have abundance of land and few men to share it, the men will all be equal." [33] Conceived as a principle of advancement based on merit, democracy met his approval as " socially progressive and profitable." Conceived as equality in acquisition and enjoyment, he thought it unintelligible in theory, and thoroughly impracticable.[34] "Industry may be republican; it can never be democratic so long as men differ in productive power and in industrial virtue." [35]

In a brilliant essay which he never published, but which was written some time before the studies of J. Allen Smith and Charles A. Beard, Sumner divined the intentions of the founding fathers in the making of the American Constitution. They feared democracy, Sumner pointed out, and attempted to set limits upon it in the federal structure; but since the whole genius of the country has inevitably been democratic, because of its inherited dogmas and its environment, the history of the United States has been one of continual warfare between the democratic temper of the people and their constitutional framework.[36]

IV

One concept of the evolutionary philosophy which Sumner borrowed from Spencer and employed with great effect in his fight against reformers was social determinism. Society, the product of centuries of gradual evolution, cannot be quickly refashioned by legislation:

The great stream of time and earthly things will sweep on just the same in spite of us. . . . Every one of us is a child of his age and cannot get out of it. He is in the stream and is swept along with it. All his science and philosophy come to him out of it. Therefore the tide will not be changed by us. It will swallow up both us and our experiments. . . . That is why it is the greatest folly of which a man can be capable to sit down with a slate and pencil to plan out a new social world.[37]

To Sumner as to Spencer, society was a superorganism, changing at geological tempo. For its emphasis upon slow change, Sumner eagerly welcomed *The Study of Sociology*. In his view, the social meddlers had been laboring under the delusion that, since there are no natural laws of the social order, they might make the world over with artificial ones; [38] but he expected that Spencer's new science would dissolve these fantasies.

With the evolutionist's characteristic scorn for all forms of meliorism and voluntarism, Sumner dismissed Upton Sinclair and his fellow socialists as puny meddlers, social quacks, who would try to break into the age-old process of societal growth at an arbitrary point and remake it in accordance with their petty desires. They started from the premise that " everybody ought to be happy " and assumed that therefore it should be possible to make everyone happy. They never asked, " In what direction is society moving? " or, " What are the mechanisms which motivate its progress? " Evolution would teach them that it is impossible to tear down overnight a social system whose roots are centuries deep in the soil of history. History would teach them that revolutions never succeed — witness the experience of France, where the Napoleonic period left essential interests much as they had been before 1789.[39]

Every system has its inevitable evils. " Poverty belongs to the struggle for existence, and we are all born into that struggle." [40] If poverty is ever to be abolished, it will be by a more energetic prosecution of the struggle, and not by social upheaval or paper plans for a new order. Human progress is at bottom moral progress, and moral progress is largely the accumulation of economic virtues. " Let every man be sober, industrious, prudent, and wise, and bring up his children to be so likewise, and poverty will be abolished in a few generations." [41]

Thus the evolutionary philosophy provided a powerful argument against legislative meddling with natural events. Sumner's conception of the proper limits of state action, although not quite so drastic as Spencer's, was severe. " At bottom there are two chief things with which government has to deal. They are the property of men and the honor of women. These it has to defend against crime." [42] Outside the field of education, where Sumner's influence was always progressive, there were few reforms proposed in America during his active years which he did not attack. In a series of essays written for the *Independent* in 1887, Sumner assailed several current reform projects as fabrications of rampant pressure groups. The Bland Silver Bill he considered an irrational compromise set up by a few public men without substantial promise of giving any real aid to debtors, silver miners, or any other part of the population. State laws limiting convict labor he damned as hasty and pointless legislation in response to partisan clamor. The Interstate Commerce Act lacked philosophy or design. The railroad question " is far wider than the scope of any proposed legislation; the railroads are interwoven with so many complex interests that legislators cannot meddle with them without doing harm to all concerned." [48] The free-silver movement he attacked with the arguments of orthodox economics.[44] " All poor laws and all eleemosynary institutions and expenditures " he scored as devices that protect persons at the expense of capital and ultimately lower the general standard of living by making it easier for the poor to live, thus increasing the number of consumers of capital while lowering incentives to its production.[45] With trade unions he was more indulgent, conceding that a strike, if conducted without violence, might be a means of testing the market conditions for labor. All the justification a strike required was success; failure was enough to condemn it. Trade unions might also be useful in maintaining the *esprit de corps* of the working class, and of keeping it informed. The conditions of labor — sanitation, ventilation, the hours of women and children — might better be controlled by the spontaneous activity of organized labor than by state enforcement.[46]

Aside from anti-imperialism, the one great dissenting impulse of his age that attracted Sumner was free trade. But, in his mind, free trade was not a reform movement; it was an intellectual axiom. Although in 1885 he wrote a short tract elaborating the classical arguments against protection (*Protectionism, The Ism That Teaches That Waste Makes Wealth*), he felt that protectionism was hardly open to dispute by enlightened men — " that it ought to be treated as other quackeries are treated." [47] Believing that tariffs, as well as other forms of government intervention in economic life, might culminate in socialism, he identified protectionism and socialism on principle, defining socialism as " any device whose aim is to save individuals from any of the difficulties or hardships of the struggle for existence and the competition of life by the intervention of ' the state.' " [48] The tariff, he admitted, never ceased to arouse his highest moral indignation. He once wrote angry protests to the newspapers because women employed in sweatshops stitching corsets for fifty cents a day had to pay a tariff on their thread.[49]

v

Intransigent against what he considered abuses of the right or left, Sumner drew fire from both sides. Upton Sinclair, in *The Goose-Step*, called him, long after his death, " a prime minister in the empire of plutocratic education "; [50] and another socialist accused him of intellectual prostitution.[51] Such critics showed little comprehension of Sumner's character or the governing motives of his mind. He was doctrinaire because his ideas were bred in his bones. He was not a business hireling, nor did he feel himself to be the spokesman of plutocracy, but rather of the middle classes. He attacked economic democracy, but he had no sympathy for plutocracy, as he understood it; he thought it responsible for political corruption and protectionist lobbies.[52] Significantly, he had praise for the Jeffersonian democracy in so far

as it practised abnegation of state power and decentraliza-
tion in government.[53]   Sumner's unforgettable " Forgotten
Man," the hero of most of his popular essays, was simply the
middle-class citizen, who, like Sumner's father, went quietly
about his business, providing for himself and his family without making demands upon the state.[54]   The crushing effect
of taxation upon such people gave Sumner his most anxious
moments and explains in part his opposition to state interventionism.[55]   It was his misfortune that this class had moved
on to the support of reform while he was still trying to fight
its cause with the intellectual weapons of Harriet Martineau
and David Ricardo.

On the rare occasions when Sumner's thought ran counter
to the established verities, he would stand his ground under
the greatest pressure.   His famous fight with President Porter
over the use of *The Study of Sociology* as a textbook might
have cost him his position at Yale, and he was quite ready
to resign.   Constantly under criticism from the press for his
outspoken stand on the tariff, he never faltered.   The New
York *Tribune,* in the course of a denunciation of his articles
on protection, once likened his manners to those of " the
cheap Tombs shyster." [56]   The Republican press and Republican alumni of Yale periodically urged his dismissal; the
demand became general when he announced his opposition
to the Spanish-American War.[57]   Although one old-fashioned
benefactor of Yale doubled his donation because Sumner's
presence had convinced him " that Yale College is a good
and safe place for the keeping and use of property and the
sustaining of civilization when endangered by ignorance, ras-
cality, demagogues, repudiationists, rebels, copperheads, communists, Butlers, strikers, protectionists, and fanatics of sundry roots and sizes," [58] Sumner was always suspect to a large
part of the community of wealth and orthodoxy because of
his independence.

Sumner's reputation has come to rest upon his *Folkways,*
and in lesser measure upon his historical writings, while his
many social-Darwinist essays have fallen into comparative
obscurity.[59]   Natural selection in the realm of ideas has taken

its toll upon his life work. The ideas for which *Folkways* is most esteemed were never reconciled with the rest of his thought. The great contribution of that work was its conception of folkways as products of "natural forces," as evolutionary growths, rather than artifacts of human purpose or wit.[60] Critics have often suggested that Sumner's denial of the intuitive character of morals, his insistence upon their historical and institutional foundations, undermined his own stand against socialists and protectionists.[61] A thoroughly consistent evolutionist, prepared to carry out the amoral and narrowly empirical approach to social change laid down in *Folkways,* would not have been so disturbed as Sumner was by the decline of laissez faire, but might have accepted it in a mellow and complaisant spirit as a new trend in the development of the mores. On the subject of laissez faire and property rights, however, Sumner was uncompromising and absolute. There is no complaisance in *Protectionism, the Ism That Teaches That Waste Makes Wealth,* no mellowness in "The Absurd Effort to Make the World Over." As a recruit from the theological life who had always been absorbed in his own Yankee culture, Sumner found the effort of a completely consistent relativism too great. It was easier for an unacclimated alien like Thorstein Veblen to treat American society with the loftiness of a cultural anthropologist. For Sumner, the marriage customs of the Wawanga and the property relations of the Dyaks were always in a separate universe of discourse from like institutions of his own culture.

As a defender of the status quo, Sumner was an effective figure in American life. Since the Revolution the dogmas of the Enlightenment had been traditional ingredients of the American faith. American social thought had been optimistic, confident of the special destiny of the country, humanitarian, democratic. Its reformers still relied upon the sanctions of natural rights. It was Sumner's function to take the leadership in a critical examination of these ideological fixtures, using as his instrument the early nineteenth-century pessimism of Ricardo and Malthus, now fortified with the tremendous prestige of Darwinism. He set himself the task of deflating the philosophical speculation of the eighteenth century with the science of the nineteenth. He tried to show his contemporaries that their optimism was a hollow defiance of the realities of social struggle, that their " natural rights " were nowhere to be found in nature, that their humanitarianism, democracy, and equality were not eternal verities, but the passing mores of a stage of social evolution. In an age of helter-skelter reforms, he tried to convince men that confidence in their ability to will and plan their destinies was unwarranted by history or biology or any of the facts of experience — that the best they could do was to bow to natural forces. Like some latter-day Calvin, he came to preach the predestination of the social order and the salvation of the economically elect through the survival of the fittest.

Chapter Four

Lester Ward

Critic

Is it true that man shall ultimately obtain the dominion of the whole world except himself?

Lester Ward

The founders of modern sociology, Comte and Spencer, were both inspired by a passion for setting the universe in order; both erected their sociological systems upon the monistic assumption that the laws of the universe at large are also applicable to human societies. One of the most impressive features of their work was the effort to arrange the subject matter of all the sciences natural and social, from astronomy to sociology, in a connected hierarchy, and to draw upon the rapidly developing physical and biological sciences for such social enlightenment as they might yield. In the spirit of this monism, Comte could speak of sociology as "Social Physics" and write, long before Darwin, of "the obvious necessity of founding sociology upon the whole of biology."[1] With the same assumptions, Walter Bagehot entitled an epochal essay in social theory *Physics and Politics,* and Herbert Spencer elaborated his analogy of the social organism, and filled his sociology with the differentiations, integrations, equilibrations, and other abstractions of his ponderous metaphysic. Spencer went so far as to deduce from the law of gravitation the intriguing sociological principle that "the attraction of cities is directly as the mass and inversely as the distance."[2]

In a peculiarly contradictory relation to this monism stood Lester Frank Ward, author of the first comprehensive sociological treatise written in the United States. Like many other youths who came of age in the early 1860's, Ward flavored his

educational diet with liberal dashes of Spencer, and admired Spencer's version of universal evolution. The monistic dogma seemed axiomatic to him. In his *Dynamic Sociology* he expressed the hope that " the universal science or
true cosmology will constitute . . . [a] great advance upon
the present heterogeneous state of science." [8] " I naturally
consider everything in its relation to the Cosmos," he wrote
near the end of his career. And of his *Pure Sociology* he once
declared: " It is more than sociology, it is cosmology." [4] The
consummation of this monism in Ward's method is readily
appreciated by the reader of *Dynamic Sociology*, who must
dig through some two hundred pages of physics, chemistry,
astronomy, biology, and embryology, before he strikes strictly
sociological data.

While Ward formally accepted the Spencerian method, his
social system, taking its shape from an entirely different pragmatic bias, was radically different in both structure and practical content. For Ward's sociology was intrinsically dualistic. Of critical importance in everything Ward wrote was
a sharp distinction between physical, or animal, purposeless
evolution and mental, human evolution decisively modified
by purposive action. By thus bifurcating the Spencerian
system, Ward sundered social principles from simple and
direct biological analogies. In his hands sociology became a
special discipline dealing with a novel and unique level of
organization. He was the first and the most formidable of
a number of thinkers who attacked the unitary assumptions
of social Darwinism and natural-law laissez-faire individualism. In time, Ward was eminently successful in impressing
his criticisms upon American sociologists. In his sphere he
served a function similar to that of instrumentalists in philosophy: he replaced an older passive determinism with a
positive body of social theory adaptable to the uses of reform.

Like many other American reformers, Ward came from a
frontier environment.[5] He was born in Joliet, Illinois, in
1841, the son of an itinerant mechanic and a clergyman's
daughter. Although Ward's youth was one of poverty and
hardship, he used the time left over from his jobs in mills,

factories, and fields to study biology and physiology, learn French, German, and Latin, and finally to qualify as a secondary-school teacher. In 1865, after two years of Civil War service, during which he suffered severe wounds at Chancellorsville, Ward entered the government service as a clerk in the Treasury Department. At twenty-six he entered evening-session college and within five years had taken diplomas in arts, law, and medicine. Much of Ward's education was self-acquired, all of it achieved through enormous sacrifice. He could never carry it lightly. Perhaps as a salve for his acute sensitivity about his humble origins, he developed a fondness for pompous Latin and Greek derivatives and sprinkled his sociology with terms like " synergy," " social karyokinesis," " tocogenesis," " anthropoteleology," and " collective telesis," called male sexual selection " andreclexis " and romantic love "ampheclexis." One of his courses at Brown University was modestly titled " A Survey of All Knowledge."

For a few years of his early government service, Ward edited, and for the most part wrote, the greater part of a journal called *The Iconoclast*. A tiny bubble on the skeptical ferment of the 1870's, full of the juvenile contentiousness of professional debunkers, it gives early evidence of Ward's complete sympathy with newer currents of thought. Later Ward continued his scientific study and in time acquired a distinguished reputation as a botanist and paleobotanist, receiving in 1883 the post of chief paleontologist in the United States Geological Survey. The same year saw the appearance of his first book, the epoch-making *Dynamic Sociology*, which he had been working on for fourteen years. In 1906, after the central conceptions of his work had been repeated and expanded in other books — *The Psychic Factors of Civilization* (1893), *Outlines of Sociology* (1898), *Pure Sociology* (1903), and *Applied Sociology* (1906) — Ward was at last called to Brown University to occupy a chair in sociology.

When Ward's *Dynamic Sociology* appeared, sociology was still in an early stage of development. While a few American universities were giving courses in a vaguely relevant subject, some using Spencer as a textbook, William Graham

Sumner was probably the only teacher who used the term " sociology " to describe a college course.[6] The materials of the subject were just emerging from courses in the " philosophy of history " and the " history of civilization." In spite of the need for a systematic treatise, the ground was ill prepared for bold theoretical innovation from an obscure government functionary, especially when he ventured to challenge the dominant Spencerian doctrines. Much to Ward's disappointment, his work was almost ignored at the start and took hold very slowly. Five years after its publication, Albion Small relates, Richard T. Ely was the only man on the otherwise alert Johns Hopkins faculty who knew it, and in 1893 Ward told him that barely five hundred copies had been sold.[7] In 1897, however, Appleton brought out a second edition, and by the turn of the century Ward was widely recognized as a first-rate figure in sociology. At least two other pioneers of American social science, Albion W. Small and Edward A. Ross, were profoundly influenced by his work, and in 1906 he was elected first president of the American Sociological Society. However, while professional sociologists learned to look to him with respect, while Small believed he saved them years of fruitless work in the arid wastes of " misconstrued evolutionism," Ward never attained a general public reputation comparable to that enjoyed by William Graham Sumner or other academic men of like stature.[8]

Ward developed his collectivism almost two decades too early to reach a fully receptive audience. Ten years before the passage of even such primitive and halting steps toward centralized control as the Interstate Commerce Act and the Sherman Act, Ward was preaching a planned society. His skepticism also restricted his influence; Christian reformers who might otherwise have been attracted to his social theory found his naturalism objectionable, and a few sympathizers urged him to be more compromising in tone.[9] Not until near the close of his career, moreover, did he hold a chair in a well-known university, and he missed the public and professional prestige that goes with a first-rate academic position.

The cumbersome prose and barbarous terminology of his formal writings, particularly *Dynamic Sociology*, which runs to fourteen hundred difficult pages, also stood in the way of a wide public reputation. Ward did, however, place readable pieces in popular journals, the most notable being a well-received series in the *Forum*.[10] Toward the end of his life, as the voices of dissent grew stronger, his thought filtered into the strategic places where the general reader was reached, and he exerted some influence upon the outlook of reform-minded groups; but partly because his proposal for a "sociocracy" never had organized adherents, his reputation faded quickly after his death in 1913. He was one of the ablest and most prescient thinkers in the history of the American mind, indeed in the history of international sociology. But it was his curious fate to be most pertinent as a thinker where he was most negative. His greatest accomplishment was as a critic of intellectual systems, once pervasive and powerful, which have long since crumbled and been forgotten. His trenchant assaults upon them, so important in his own day to the liberation of American thought, have been forgotten with them.

II

Ward had felt all too keenly the sting of his lower-class origin, and the aristocratic innuendoes of social Darwinism as it found expression in the 1870's and 1880's offended his democratic sensibilities. To the end of his life he remembered how, as a child in public school, he had felt keen satisfaction whenever the ragged boys of his own class were able to beat the sons of rich men in scholarship.[11] If his childhood experiences had something to do with encouraging his faith in the latent intellectual capacities of the masses, Ward's long experience in government agencies may have encouraged him to oppose the Spencerian distrust of government. As early as 1877, after a few years of service in the Bureau of Statistics, he wrote two articles for the Washington *National Union* in which he explored the possibilities of government statistics as a basis for legislation, arguing that if the laws of social events could be statistically formulated they could be used as data for "scientific lawmaking." [12]

Within the next few years Ward's political interests became increasingly urgent. He was already well along in the writing of *Dynamic Sociology*. In a paper read in 1881 before the Anthropological Society of Washington, he made a headlong assault upon the fundamental premises of the prevailing laissez-faire philosophy. Here he set forth in striking fashion the ideas to which he was to devote his later life. Pointing out that the prevailing trend toward government intervention in social affairs was utterly incompatible with existing social theory, Ward predicted, with no little prescience, that a crisis in social opinion would soon be precipitated.

> The Cobden Club and other "Free Trade" societies are scattering tracts with a liberal hand, in the hope of stemming the tide. Victor Boehmert warns, Augustus Mongredien shouts, and Herbert Spencer thunders. What is the result? Germany answers by purchasing private railroads and enacting a high protective tariff. France answers by decreeing the construction of eleven thousand miles of Government railroad, and offering a bounty to French ship-owners. England answers by a compulsory education act, by Government purchase of the telegraph, and by a judicial decision laying claim to the telephone. America answers by an inter-state railroad bill, a national education bill, and a sweeping *plebiscite* in favor of protection to home manufacturers. The whole world has caught the contagion, and all nations are adopting measures of positive legislation.[13]

It was time, Ward continued, for scholars to stop decrying this irresistible trend toward legislative intervention and settle down to a serious study of what was going on. The natural-law and laissez-faire dogmas had been useful intellectual devices in the days when society was being freed from monarchical and oligarchical rule. It was natural enough to oppose governmental interference when government was in the hands of autocrats, but it is folly to cling to this opposition in an age of representative government when the popular will can be exerted through legislative action. The assumptions are obsolete. "There is no necessary harmony between natural law and human advantage" The laws of trade result in enormous inequalities in the distribution of wealth, which are founded in accidents of birth or strokes of low cunning rather than superior intelligence or industry.

Nor is natural law a barrier against monopolies. The classical theory says that competition keeps prices down, but often competition "multiplies the number of shops far beyond the necessity, each of which must profit by exchange, and in order to do this all must sell dearer than would otherwise be necessary." This is particularly true of the distributive industries. In other lines competition breeds huge corporate organizations with dangerously broad powers. To break them up would be to destroy "the legitimate product of natural law," the "integrated organisms of social evolution." The only constructive alternative is government regulation in the interest of society at large.[14] Historic attempts at government regulation or management have not been the disasters that individualists charge. Witness the telegraph in Great Britain and the railroad systems of Germany and Belgium. The sphere of social control has been gradually expanding in the history of civilization, but

. . . for more than a century the English school of negative economists has devoted itself to the task of checking this advance. The *laissez faire* school has entrenched itself behind the fortifications of science, and while declaring with truth that social phenomena are, like physical phenomena, uniform and governed by laws, they have accompanied this by the false declaration and *non sequitur* that neither physical nor social phenomena are capable of human control; the fact being that all the practical benefits of science are the result of man's control of natural forces and phenomena which would otherwise have run to waste or operated as enemies to human progress. The opposing positive school of economists simply demands an opportunity to utilize the social forces for human advantage in precisely the same manner as the physical forces have been utilized. It is only through the artificial control of natural phenomena that science is made to minister to human needs; and if social laws are really analogous to physical laws, there is no reason why social science may not receive practical applications such as have been given to physical science.[15]

In an article on "The Scientific Basis of Positive Political Economy" (1881) Ward continued his assault upon natural làw in social theory. By human standards, he asserted, nature itself is uneconomic. That its process proves " the least economic of all conceivable processes " is concealed only by the vastness of the scale on which nature operates and the absolute magnitude of its results. Some of the lower organisms give off as many as a billion ova: only a few develop into maturity, while the rest succumb in the resulting struggle for survival. The waste of reproductive powers is fantastic. Haphazard human strife, particularly in the form of industrial competition, is similarly wasteful. Here Ward distinguished between telic phenomena — those governed by human will and purpose — and genetic phenomena, the results of blind natural forces. In the face of the immense superiority of the telic over the genetic, the artificial over the natural, the persistent natural-law enthusiasm of laissez-faire theorists is like the nature-worship of Rousseauian romanticism, or, worse still, of primitive religion. The evolutionary view of nature as being in some way inherently beneficent is sheer mysticism.[16] Man's task is not to imitate the laws of nature, but to observe them, appropriate them, direct them.

Just as there are two kinds of dynamic processes, so are there two distinct kinds of economics — the animal economics of life and the human economics of mind. Animal economics, the survival of the fittest in the struggle for existence, results from the multiplication of organisms beyond the means of subsistence. Nature produces organisms in superabundance and relies upon the wind, water, birds, and animals to sow her seed. A rational being, on the other hand, prepares the ground, eliminates weeds, drills holes, and plants at proper intervals; this is the way of human economics. While environment transforms the animal, man transforms the environment.

Competition actually prevents the most fit from surviving. Rational economics not only saves resources but produces superior organisms. The best evidence for this is that whenever competition is wholly removed, as it is when man artificially cultivates a particular form of life, that form immediately makes great strides and soon outstrips those depending upon competition for their progress. Hence the superior quality of fruit trees, cereals, domestic cattle. Even in its most rational form, competition is prodigiously wasteful. Witness the social waste involved in advertising, a good example of " the modified form of animal cunning " which is the hallmark of business shrewdness. Finally, with the gusto of a debater making his clinching point, Ward argued that laissez faire actually destroys whatever value competition might have in human affairs; for since complete laissez faire allows combination and finally monopoly, free competition can be secure only through some measure of regulation.[17]

In his *Dynamic Sociology*, which was inspired by " a growing sense of the essential sterility of all that has thus far been done in the domain of social science," and designed as a reply to those who " conclude that Nature's ways should be man's ways," [18] Ward massed all his arguments against natural law and expanded his plea for teleological progress. While he always scorned the name of reformer and insisted that he was a social scientist, *Dynamic Sociology* was essentially an argument for socially organized and guided reform — or, as Ward preferred to call it, " the improvement of society by cold calculation " — which, he believed, was destined to replace the hitherto automatic processes of social change.[19] When Ward first began work on *Dynamic Sociology*, he planned to call it *The Great Panacea*.

Ward made one concession to biological theory: he agreed that man has been brought to his present stage of development by natural selection, of which his intellect is the supreme product; but man cannot consider himself finally superior to other animals until he supplants genetic with telic progress by applying his intellect to his own improvement.[20] Social progress consists in an increase in the aggregate enjoyment throughout a society and a decrease in the aggregate suffering.

Thus far, social progress has in a certain awkward manner taken care of itself, but in the near future it will have to be cared for. To do this and maintain the dynamic condition against all the hostile forces which thicken with every new advance, is the real problem of Sociology considered as an applied science.[21]

In his second volume Ward stressed the importance of feeling in social dynamics. Feelings, he insisted, are the basic component of mind; the intellect has been evolved as a guide to the feelings. The social mind, a generalization or composite of individual minds, is made up of the social intellect and social feelings. The unrestrained working out of feelings results in conflict and destruction; but intellect can guide the feelings into constructive channels by setting down laws and ideals. Intellect, in its growth, finally becomes capable of formulating ideals for social as well as individual guidance.

Those actions which bring progress, which Ward called "dynamic actions," can be performed only by creating a state of "dynamic opinion" in which the social intellect is equipped for its guiding function.[22] If a whole society is to embark upon a dynamic action, its people must be prepared and equipped through the broadest possible diffusion of knowledge.

Intelligence, hitherto a growth, is destined to become a manufacture. The knowledge of experience is, so to speak, a genetic product; that of education is a teleological product. The origination and distribution of knowledge can no longer be left to chance and to nature. They are to be systematized and erected into true arts. Knowledge artificially acquired is still real knowledge, and the stock of all men must always consist chiefly of such knowledge. The artificial supply of knowledge is as much more copious than the natural as is the artificial supply of food more abundant than the natural supply.[23]

For Ward education was more than a device for social engineering; it was also a leveling instrument, a means of bringing opportunity to humble people and enabling them to use their talents.[24] Greatly impressed from his childhood by the vast difference between the educated and uneducated, Ward was never able to believe that this chasm, which he himself had bridged, could be attributed to differences in native capacities. His passionate emphasis on education sprang from his own personal triumph.[25]

Because he believed in education as a long-term instrument for the improvement of mankind, Ward was reluctant to surrender the Lamarckian and Spencerian notion of the transmission of acquired characteristics. This idea, which Darwin had accepted, but had not at first made an integral part of his theory of evolution, Ward considered a necessary ingredient in his optimistic sociology. On a number of occasions he clashed with neo-Darwinians like Weismann. In a significant article published in the *Forum* in 1891 on " The Transmission of Culture," he granted that acquired knowledge itself cannot be transmitted by heredity, but insisted that the *capacity* to acquire knowledge is another matter. Certain arts and talents which apparently run in family lines cannot be accounted for by the theory of natural selection because these talents have no value in the struggle for survival; natural selection has no explanation for the persistence of such talents from generation to generation. The persistence of talents can best be explained by assuming that part of what man gains by the exercise of mental faculties in a specific pursuit may be handed down to become part of the heritage of the race. If Weismann's followers are right, and there is no such inheritance, then " education has no value for the future of mankind and its benefits are confined exclusively to the generation receiving it." The facts of history and personal observation sustain the general popular belief in such use-inheritance, Ward concluded, and until the matter should be definitely decided by science, it would be well to " hug the delusion." [26]

III

Ward is sometimes classified among the social Darwinists because his later theory was influenced by the conflict school of sociologists, represented most prominently by two continental writers, Ludwig Gumplowicz and Gustav Ratzenhofer. By 1903 Ward had become well acquainted with their works, and was so impressed by their interpretation of the origins of race struggle, which he called " the most important contribution thus far made to the science of sociology," [27] that he based a small part of his *Pure Sociology* upon it. There he attributed the origin of organized society to the conquest of one race by another. Caste systems had developed out of such conquest, and society had then passed successively through five stages: the mitigation of caste coupled with the survival of inequalities; the consolidation of relationships through the growth of law; the origin of the state; the gradual cementing of the groups into a homogeneous people; and, finally, the development of patriotism and the national form of social organization.[28]

Progress has frequently resulted from the forcible fusion of unlike elements. As much as one may deplore the horrors of war, it has been a necessary condition of race progress in the past, and the conquest of backward races is inevitable in the future.[29] In advanced societies, rational and peaceful forms of social assimilation may supersede the genetic and violent method of the past. It is possible that a friendly pacific age is about to dawn — just as Spencer's militant type of society gives way to the industrial — but it is doubtful that the world has yet reached the point at which war ceases. Whether the cessation of conflict would even be desirable was an open question to Ward.[30]

His adherence in these respects to the conflict school did not in the least alter the fundamental structure of Ward's melioristic sociology. He saw no difficulty — although there were in fact grave difficulties — in reconciling the conflict theory with his collectivism; he even succeeded in converting Gumplowicz to his own cheerful point of view.[31] The ideas of the conflict school found but a small and transient place in Ward's work, and the theory of his later years is not otherwise markedly different from that of 1883. Throughout the greater part of all his writing, his aim was to destroy the tradition of biological sociology.

A persistent feature of Ward's sociology was his running argument with the paralyzing optimism of the Spencerians and the equally paralyzing pessimism of the Malthusians. He regarded both the Malthus-Ricardo-Darwin lineage of pessimism and Spencerian optimism as upper-class apologies for social oppression and misery.[32]   Malthus' theory, he objected, does not apply to *genus homo*.   Malthus had uncovered a fundamental law of biology, said Ward, but in attaching it to mankind he had applied it to the one animal for which it had no validity.   Darwin had had the genius to illuminate the processes of the whole organic world by applying Malthusianism fruitfully to animals and plants.

Notwithstanding the failure of Malthusianism at all points, the impression prevailed, and still prevails, that it is a fundamental law of society, and the current sociology is based upon it. . . .   The fact is that man and society are not, except in a very limited sense, under the influence of the great dynamic laws that control the rest of the animal world. . . .   If we call biologic processes natural, we must call social processes artificial.   The fundamental principle of biology is natural selection, that of sociology is artificial selection.   The survival of the fittest is simply the survival of the strong, which implies and would better be called the destruction of the weak.   If nature progresses through the destruction of the weak, man progresses through the protection of the weak.[33]

Ward did not hesitate to cross swords with Spencer or Spencer's American disciples, Sumner and Giddings.   Probably the most unfavorable review received by Sumner's *What Social Classes Owe to Each Other* was written by Ward for the New York periodical *Man*.   This book, said Ward, was the " final wail " of the laissez-faire writers.   It would do more good than harm because it was so extreme as to be a caricature of individualism.

The whole book is based on the fundamental error that the favors of this world are distributed entirely according to merit.   Poverty is only a proof of indolence and vice.   Wealth simply shows the industry and virtue of the possessors.   The very most is made of Malthusianism, and human activities are degraded to a complete level with those of animals.   Those who have survived simply prove their fitness to survive; and the fact which all biologists understand, viz., that fitness to survive is something wholly distinct from real superiority, is, of course, ignored by the author because he is not a biologist, as all sociologists should be.[34]

In an extended polemic against " The Political Ethics of Herbert Spencer," [35] Ward skillfully selected passages from

Spencer in which he relies upon the beneficence of traders
to refrain from making merciless bargains and excess profits;
passages in which he defends private control of sewage systems by suggesting that payments to sewage companies can
be enforced by threats to turn off the drainage facilities of
recalcitrant householders; passages in which he speaks of the
unemployed as " good-for-nothings " and of a trade union as
" a permanent body of tramps "; passages in which he ex-
presses an aristocratic disdain for democratic processes; and
similar individualistic extremisms. He went on to exploit
the contradiction between Spencer's individualism and his
organismic view of society. If the state, the supreme organ
of integration, is to have practically no function, Ward asked,
what becomes of Spencer's increasing integration as a criterion of progress? The logical outcome of the social organism is not extreme individualism but extreme centralization. " The strongest advocate of state control, the most
extreme socialist, would shrink from the contemplation of
any such absolutism as that exercised by the central ganglion
of even the lowest of the recognized Metazoa." [36] The organismic analogy is sound only when it refers to the psychic
aspects of society, and even on this level it logically implies an
extension of social control, because a government is the servant of the popular will in the same way that the brain is a
servant of the animal's will.[37]

Still another foible of the Spencerians was their loaded
definition of the term " natural," which they rather incon-
sistently used not to describe whatever phenomena they
might find but only those phenomena of which they could
approve. In fact, however, the inertia of society and its failure to respond at once to the pressure of change " gives rise
to social reformers who are legitimate and necessary, nay,
natural products of every country and age, and the ignoring
of this fact by conservative writers who lay so great stress on
the word *natural* is one of the amusing absurdities of the
present period." [38]

Rejecting the premises of classical individualism, Ward
was impelled to strike out on untried ideological lines, to

develop an approach to social theory in terms of psychology and institutions rather than biology and individuals. Like most other professional biologists he was little impressed by the facile analogies between nature and society that pleased the apologists of the competitive order. Unable to find in society the crude processes he saw at work in nature, Ward evolved a twofold criticism of social Darwinism. He first debunked nature itself, displayed its wastefulness, and tore it from the high place it occupied in the popular mind. Then, by showing how the emerging human mind was able to mold the narrow genetic processes of nature into vastly different forms, Ward demolished the central feature of the monistic dogma — the continuity between process in nature and process in society.

Darwinism, with its emphasis upon gradual change over geological periods and its interpretation of change as a result of "accidental" variations, had appeared to banish teleology from the animal world, and thus, for those who worked in the shadow of the monistic dogma, banished it from the human world also. If there were no larger purpose, no cosmic guiding hand behind the emergence of higher species, if evolution was a planless outcome of random variations, purposefulness had no place in the universe, and societies must grow and change as aimlessly as the rest of life. To Ward, however, it seemed that the reaction from teleology had gone too far. If there is no cosmic purpose, there is at least human purpose, which has already given man a special place in nature and may yet, if he wills it, give organization and direction to his social life. Purposeful activity must henceforth be recognized as a proper function not only of the individual but of a whole society.

IV

Always cosmopolitan in his interests, Ward was much concerned from the first to interpret to Americans the lessons of European thought and experience on the subject of state intervention. Apart from his own insight as a government employee, he was impressed by the extension of state activity abroad, particularly in government ownership or regulation of railroads as it could be observed in Germany, France, Belgium, and England.[39]   When he compared European methods with the American practice of private management, it was to the detriment of the latter.[40]   He was also influenced against laissez faire by the critical attitude of Comte, whom he greatly admired.[41]

This is not to say, of course, that Ward was merely another nationalist in economics. His advocacy of state management was prompted by a lower-class bias. He seems to have considered himself a lobbyist for the people in academic forums. His opposition to the biological argument for individualism stemmed from his democratic faith; his rejection of Sumner and Spencer was partly motivated by his sense of their aristocratic preferences. Like Veblen, Ward felt a certain personal alienation from the dominant characters and opinions of American intellectual life, which doubtless quickened his championship of the underdog. He once complained of the " capitalistic censorship " at the University of Chicago. And during the campaign of 1896 he wrote to E. A. Ross, who was suffering for supporting Bryan, " I would probably go further toward populism than you. No one is more anxious to throttle the money power," adding only that he considered free silver a poor social remedy and that he had no desire to go through another monetary inflation like the one of his youth.[42]

Ward made a revealing statement of his social bias at an American Sociological Society meeting in 1906 during a discussion of " Social Darwinism." A previous speaker had presented a social-Darwinist thesis advocating careful elimination of the unfit and dependent, chiefly by eugenic methods. In reply Ward branded the doctrine presented as " the most complete example of the oligocentric world-view which is coming to prevail in the higher classes of society and would center the entire attention of the whole world upon an almost infinitesimal fraction of the human race and ignore the rest." He would not be contented, Ward continued, to work in so small a field as the education and preservation of a select few of the higher classes. "I want a field that shall be broad enough to embrace the whole human race, and I would take no interest in sociology if I did not regard it as constituting such a field." For an indefinite period to come, society would be recruited from the base and would be compelled to assimilate a mass of crude material from the bottom. His opponents might conclude from this that " society is doomed to hopeless degeneracy." Yet it is possible to take another view:

> . . . the only consolation, the only hope, lies in the truth . . . that so far as the native capacity, the potential quality, the "promise and potency," of a higher life are concerned, those swarming, spawning millions, the bottom layer of society, the proletariat, the working class, the "hewers of wood and drawers of water," nay, even the denizens of the slums — that all these are by nature the peers of the boasted "aristocracy of brains" that now dominates society and looks down upon them, and the equals in all but privilege of the most enlightened teachers of eugenics.[43]

Although he was a forerunner of social planning, a champion of the masses, applauded and used by those socialists who read his works, Ward himself was no socialist. He was singularly uninterested in the Marxian tradition. He believed that he had a workable alternative to socialism and individualism which, borrowing from Comte, he called " sociocracy," or the planned control of society by society as a whole. Under sociocracy, purposeful social activity, or " collective telesis," could be harmonized with individual self-interest by means of " attractive legislation " designed to release the springs of human action for socially beneficial deeds by positive rather than negative and compulsory devices. Where individualism has created artificial inequalities, sociocracy would abolish them; and while socialism seeks to create artificial equalities, sociocracy would recognize inequalities that are natural. A sociocratic world would distribute its favors according to merit, as individualists demand, but by equalizing opportunity for all it would eliminate advantages now possessed by those with undeserved power, accidental position or wealth, or antisocial cunning.[44]

In his anticipation of social planning and his historical perspective on the limitations of laissez faire, as well as in his campaign against biological sociology, Ward did much to relieve American thinking from its uncritical preoccupation with the conservative uses of nineteenth-century science. In social psychology he helped his followers to arrive at a better understanding of the importance of feeling in human motivation. When he attempted to offer positive programs, he was vulnerable to criticism for his naïve faith in the possibilities of education for social reconstruction and for the vagueness of some of his reform proposals. Philosophically he was not the most consistent or the most sophisticated critic of monistic thinking. On the abstract level he left much to be done by the pragmatists. While Ward's dualism of the genetic and the telic was in effect a departure from what William James called the " block-universe " of Spencer, the Spencerian virus remained in his blood. In the midst of his attack upon the sociological nature-worshipers he could lapse into their own language by characterizing large combinations as products of the natural order; and he once wrote that collective telesis alone could " place society once more in the free current of natural law." [45] If he recognized the breach in his system at all, he simply covered it by saying that telic behavior is a genetic product. For one who emphasized so constantly the unique and artificial character of social organization and social processes, it was an odd inconsistency to deck out his sociology with physics, chemistry, and biology, and to set it in the framework of a cosmological system.

However unfinished Ward's critique was in a technical sense, it was nevertheless a bold pioneering stroke. He suffered much undeserved neglect partly for the very reason that he was so far in advance of the rest of his generation. " You were not only ahead of us in point of time," Albion Small wrote to him in 1903,[46] " but we all know that you are head, shoulders and hips above us in many respects scientifically. You are Gulliver among the Lilliputians."

Chapter Five

Evolution, Ethics, and Society

I have received in a Manchester newspaper rather a good squib, showing that I have proved "might is right," and therefore that Napoleon is right, and every cheating tradesman is also right.

CHARLES DARWIN to SIR CHARLES LYELL

The age in which Spencer, Sumner, and Ward formulated their philosophies was one of great intellectual insecurity. While, as we have seen, many men were uncertain how much of their religion would be left standing after natural selection had been fully accepted, others were quite as troubled by questions about what Darwinism would mean for the moral life. Spencer and the evolutionary anthropologists promised them that it would mean progress, perhaps perfection.[1] The Malthusian element in Darwinism, however, pointed to an endless struggle for existence regulated by no sanction more exalted than mere survival. While some expected a new and higher morality, others feared a complete collapse of moral standards.

Senator Gore, one of the characters in Henry Adams' novel *Democracy* (1880), which was set against the dissolute and money-mad atmosphere of Washington in the Gilded Age, expressed the essential aimlessness and sterility of what many men feared would be the dominant values of the future:

But I have faith; not perhaps in the old dogmas, but in the new ones; faith in human nature; faith in science; faith in the survival of the fittest. Let us be true to our times, Mrs. Lee! If our age is to be beaten, let us die in the ranks. If it is to be victorious, let us be the first to lead the column. Anyway, let us not be skulkers or grumblers.[2]

Men with a deeper sense for traditional ideals hoped for more than this. Did Darwinism really justify brutal self-assertion, the neglect of the weak and the poor, the abandonment of philanthropic enterprise? Did it mean that progress must be dependent upon the ruthless elimination of the unfit, in an expanding population forever pressing upon the bounds of subsistence?

In a nation trained in Christian ethics and fortified by a democratic and humanitarian heritage, such a Nietzschean ransvaluation of values was out of the question. Spencer's econciliation of evolution and idealism, with its forecast of a transition from militancy to peace and from egoism to altruism was the commonest answer. Yet Spencer often spoke in rude selectionist language which could satisfy few who were not uncompromising defenders of a strictly competitive order or who were not willing to make drastic concessions to a naturalistic ethic, bare of all the warm and familiar theological sanctions. In *The Principles of Sociology,* he declared:

> Not simply do we see that in the competition among individuals of the same kind, survival of the fittest has from the beginning furthered production of a higher type; but we see that to the unceasing warfare between species is mainly due both growth and organization. Without universal conflict there would have been no development of the active powers.[3]

In the light of all this talk about "unceasing warfare" and "universal conflict," what was the value to those interested in the here and now of Spencer's promise of a remote social Nirvana? One philanthropist asked:

> Would not mankind take chloroform if they had no future but Spencer's? No individual continuance, no God, no superior powers, only evolution working towards a benevolent society here and perfection on earth, with great doubt whether it could succeed, and, if it succeeded, whether the end would pay.[4]

"Herbert Spencer's ethics will certainly be the final ethics," wrote another critic, "but the question does press itself upon us, what is to be the ethics for the time now present and passing?"[5] "What are we to do," queried James McCosh, "with our reading youth entering upon life who are told in scientific lectures and journals that the old sanctions of morality are all undermined? " [6]

In 1879 the *Atlantic Monthly* published an essay by Goldwin Smith with the significant title " The Prospect of a Moral Interregnum," which faced the troublesome questions raised by naturalism. Religion, Smith believed, had always been the foundation for the western moral code; and it would be idle for positivists and agnostics to imagine that while Christianity was being destroyed by evolution the humane values of Christian ethics would persist. Ultimately, he conceded, an ethic based upon science might be worked out, but for the present there would be a moral interregnum, similar to those which had occurred in past times of crisis. There had been such an interregnum in the Hellenic world after the collapse of its religion brought about by scientific speculation; there had been another in the Roman world before the coming of Christianity gave it a new moral basis; a third collapse in western Europe following the Renaissance had produced the age of the Borgias and Machiavelli, the Guises and the Tudors; finally, Puritanism in England and the Counter Reformation in the Catholic Church had reintroduced moral stability. At present another religious collapse is under way:

> What then, we ask, is likely to be the effect of this revolution on morality? Some effect it can hardly fail to have. Evolution is force, the struggle for existence is force, natural selection is force. . . . But what will become of the brotherhood of man and of the very idea of humanity? [7]

What would keep the stronger races from preying on the weak? (Smith had heard of an imperialist who said, " The first business of a colonist is to clear the country of wild beasts, and the most noxious of all wild beasts is the wild man.") Or, if a tyrant should seize the reins of power in any of the great states, what could be said against him, consistently, under the survival doctrine? (Had not Napoleon been selected for survival?) What would happen to nineteenth-century humanitarianism? How were the passions of social conflict to be abated? To these questions Smith had no answer, but he was sure that the impending crisis in morals would bring with it a crisis in politics and the social order.

Other writers concerned themselves with more concrete issues. Francis Bowen, Professor of Moral Philosophy at Harvard, who could never overcome his religious hostility to Darwin, probably expressed the attitude of many old-school Christian conservatives when he attempted to discredit Darwinism by accentuating its dire social consequences. Familiar with the Malthusian genealogy of natural selection, Bowen linked the two as twin errors. Malthusianism had become popular in England, he pointed out, because it had counteracted the revolutionary ideas of men like Godwin; but it had also been used to relieve the rich of responsibility for the sufferings of the poor. Malthus had been proved wrong by the course of events; and just when his theory was dying out in political economy it received fresh support from Darwinian biology. The same arguments against the theory still hold good; for the social process is the opposite of the Darwinian process. It is undeniable that the lower classes are more fertile than the upper, that the unfittest rather than the fittest survive. Thus it is the existence of higher, not lower, forms that is imperiled in the social process. The solution to this can come only from persons of wealth, culture, and refinement, who must violate the canons of Malthus and propagate more freely to promote civilization. Wherever the Darwin-Malthus system is applied its consequences are bad: in sociology a hard-hearted indifference to the sufferings of the poor; in religion, atheism; in philosophy the dark wastes of German pessimism, and a contempt for the value of human life which, like Stoicism in Rome, presages social catastrophe.[8]

Comparable in its social conservatism, but more congenial to the scientific spirit, was the view of another writer who predicted a great conflict between what he termed " the sympathetic and the scientific theories of government." The sympathetic party is all for alleviating the condition of the working class by social legislation. No such philanthropic softness is really needed in the United States, where only natural incapacity prevents a man from becoming a capitalist. The masses cannot be artificially saved from their own incompetence without social disaster. American society, under the influence of the philanthropists of the sympathetic party, is being deluged by a flood of immigrants and dragged down by an increasing proportion of incapables. The scientific party would "defend the principle of competition, conformity to the law of supply and demand, and a fair field for the experiment of the survival of the fittest." [9]

The doctrines of the "scientific party" were similar to those of the comfortable set whose social prejudices William Dean Howells so coldly examined in *A Traveler from Altruria* (1894). To Mr. Homos, who was appalled at the apparent rigidity of class barriers in American society, the American explained:

"The divisions among us are rather a process of natural selection. You will see, as you get better acquainted with the workings of our institutions, that there are no arbitrary distinctions here, but the fitness of the work for the man and the man for the work determines the social rank that each one holds. . . ."
I added: "You know we are a sort of fatalists here in America. We are great believers in the doctrine that it will all come out right in the end."
"Ah, I don't wonder at that," said the Altrurian, "if the process of natural selection works so perfectly among you as you say." [10]

Within the "scientific party" itself some had doubts about the possibilities of progress. When an essayist in the *Galaxy* protested against the general blind faith in machines, inventions, and popular reforms, and argued that the panaceas of enthusiasts were impotent in the face of population pressures, he was answered with a blast of evolutionary optimism by George Cary Eggleston in the columns of *Appleton's*. There is no need, said Eggleston, to bewail the pressure of population or to limit its growth. The crowding of the world, by stimulating industry and forcing men to develop their capacities, by crushing the unfit, " by casting out the unworthy and raising the worthy to prosperity and power," acts as the greatest motive power of progress.

A more humanitarian attitude was voiced by the eminent geologist Nathaniel S. Shaler, a scientist in the " sympathetic party," who questioned the value of numbers in society. Shaler pointed out that it was characteristic of the higher species to be less wasteful in having progeny, that civilization replaced natural selection with selection by intelligence. If natural selection were really operating to full effect in civilization, Shaler would admit the desirability of an increase in population, but in fact humanity dictates the preservation of all, weak or strong, who come into existence, and even modern warfare selects for survival the weak, cowardly, and superannuated and destroys the fit. It would then be better to rely on education to supply the select few that nature would produce in a more wasteful way. Education demands a high standard of comfort, which in turn demands the " limitation of reproduction to the true needs of the race." [11]

In such terms as these, unsettled questions made their way into popular forums. Readers who turned to serious books between 1871 and 1900 found much provocative discussion of the meaning of Darwinism for ethics, politics, and social affairs. There were others besides Sumner and Spencer who had a powerful effect upon American intellectual life. One, John Fiske, was a native son, but most were English. Walter Bagehot, Huxley, Henry Drummond, Benjamin Kidd, William Mallock — such men were as much leaders in American thought as almost any American writer. At least one continental thinker also, Prince Peter Kropotkin, received a favorable hearing. The merits of their contributions were unequal, but all were listened to with respect.

II

Darwin himself offered somewhat confused counsel on the ethical implications of his own discoveries. In the light of his discussion of the moral sense and the role of sympathy in evolution, it is not surprising to find him somewhat hurt at the suggestion that he had proved that might is right. He little suspected that he was fated to be an intellectual Pandora; for, however dismal the Malthusian logic behind his system, it was filtered through his own tender moral sensibilities. True, *The Origin of Species* was Hobbesian in spirit, and Darwin's remarks on " Natural Selection as Affecting Civilised Nations " in *The Descent of Man* were at points reminiscent of the harshest portions of Spencer's *Social Statics:*

> We civilised men . . . do our utmost to check the process of elimination; we build asylums for the imbecile, the maimed, and the sick; we institute poor-laws; and our medical men exert their utmost skill to save the life of every one to the last moment. . . . Thus the weak members of civilised society propagate their kind. No one who has attended to the breeding of domestic animals will doubt that this must be highly injurious to the race of man.[12]

Yet this was not characteristic of Darwin's moral sentiments, for he went on to say that a ruthless policy of elimination would betray " the noblest part of our nature," which is itself securely founded in the social instincts. We must therefore bear with the evil effects of the survival and propagation of the weak, and rest our hopes on the fact that " the weaker and inferior members of society do not marry so freely as the sound." He also advocated that all who cannot spare their children abject poverty should refrain from marriage; here again he lapsed into Malthusianism with the statement that the prudent should not shirk their duty of maintaining population, for it is through the pressure of population and the consequent struggles that man has advanced and will continue to advance.[13]

If there were, in Darwin's writings, texts for rugged individualists and ruthless imperialists, those who stood for social solidarity and fraternity could, however, match them text for text with some to spare. Darwin devoted many pages of *The Descent of Man* to the sociality of man and the origins of his moral sense. He believed that primeval men and their apelike progenitors, along with many lower animals, were probably social in their habits, that remote primitives practised division of labor, and that man's social habits have been of enormous importance in his survival. " Selfish and contentious people will not cohere," he wrote, "and without coherence nothing can be effected." He believed man's moral sense to be an inevitable outgrowth of his social instincts and habits, and a critical factor in group survival. The pressure of group opinion and the moral effect of family affections he ranked with intelligent self-interest as biological foundations of moral behavior.[14] It was little wonder that when Kropotkin wrote his *Mutual Aid* he claimed Darwin as a predecessor and blamed others for putting a Hobbesian interpretation on Darwin's theory.[15]

Two years after *The Descent of Man* appeared the first significant work of biologically derived social speculation to break Spencer's monopoly in that field — Walter Bagehot's *Physics and Politics,* more aptly described in its subtitle, *Thoughts on the Application of the Principles of " Natural Selection " and " Inheritance " to Political Society.* Published in Youmans's International Scientific Series, Bagehot's book met an immediate favorable reception in this country, and did much to encourage social interpretation along biological lines. Bagehot attempted to reconstruct the pattern of growth of political civilization in the manner of evolutionary ethnologists like Lubbock and Tylor, from whom he drew some of his data.

Bagehot did not try to explain the circumstances under which law and political institutions originated. " But when once politics were begun, there is no difficulty in explaining why they lasted. Whatever may be said against the principle of ' natural selection ' in other departments, there is no doubt of its predominance in early human history. The strongest killed out the weakest as they could." Since any form of political organization was superior to chaos, an aggregation of families having political leadership and some legal custom would rapidly conquer those that did not. The caliber of early political organization was less important than the fact that it was there at all; its function was to create a " cake of custom " which would bind men together, holding them, to be sure, in whatever place in the social order birth had given them — for organization originates in a regime of status and only long afterward evolves into a regime of contract. The second step, after organization, is the molding of national character. This came about through the unconscious imitation of a chance " variation " displayed by one or two outstanding individuals. The national character is simply the naturally selected parish character, just as the national speech is the successful parish dialect.

Progress, habitually thought of as a normal fact in human society, is actually a rare occurrence among peoples: the ancients had no such conception, nor do orientals; and savages do not improve. The phenomenon occurs only in a few nations of European origin. Some nations progress while others stagnate, because under all circumstances the strongest prevail over others; and the strongest are, " in certain marked peculiarities," the best. Within each nation the most appealing character, usually the best, prevails; and in the now dominant western part of the world these competitions between nations and character types have been intensified by " intrinsic forces." Of the existence of progress in the military art there can be no doubt, nor of its corollary that the most advanced will destroy the weaker, that the more compact will eliminate the scattered, and that the more civilized are the more compact. An advance in civilization is thus a military advantage. Backward civilizations, being more rigid in the structure of their law and custom, " kill out varieties at birth," but progress depends upon the emergence of varieties. ' Progress is only possible in those happy cases where the force of legality has gone far enough to bind the nation together, but not far enough to kill out all varieties and destroy nature's perpetual tendency to change." Early societies were in a grave dilemma: they needed custom to survive, but unless it was sufficiently flexible to admit variations they were frozen in their ancient mold. Modern societies, living in an age of discussion rather than rigid custom, have found a means of reconciling order with progress.[16]

Darwin's task of finding natural roots for man's moral feelings and for the sympathy that underlies persistent social coöperation was taken up by John Fiske in his *Outlines of Cosmic Philosophy* (1874) and *The Meaning of Infancy* (1883). After reading Alfred Wallace's account of his observations in the Malay Archipelago, Fiske had been struck by the thought that one thing that distinguishes the human race from the other mammals is the very long duration of its infancy. In general there is a correlation between the complexity of a species' potential behavior and the proportion of its behavior that is acquired by learning after birth. The human infant acquires the smallest proportion of its ultimate capacities during gestation; it is born less developed than the young of other species, and must undergo a long plastic period in which it learns the ways of its race. What makes the human species progressive, Fiske reasoned, is the fact that the infant does not come into the world with its capacities "all cut and dried," but on the contrary must learn slowly and is therefore able to learn an infinitely wider range of behavior. The necessity of seeing infants through this long period prolongs the years of maternal affection and care and tends to keep father, mother, and child together — in short, to found the stable family and ultimately the clan organization, the first step toward civil society. From being merely gregarious, man becomes social.

Once the clan is organized, natural selection intervenes to maintain it; for those clans in which the primeval selfish instincts were most effectively subordinated to the needs of the group would prevail in the struggle for life. In this way the first germs of altruism and morality, manifest in the mother's care of the infant, become generalized into wider and wider social bonds until they form sympathies broad enough to support the communal life of civilized man as he is now known. The moral sense has its foundation in the primitive biological unit, the family, and the social coöperation and solidarity of men is nothing if not natural.[17]

Fiske's philosophy attempted to give to the higher ethical impulses a direct root in the evolutionary process. A somewhat different — and, to most of his contemporaries, a less satisfactory — note of moral reassurance was struck by T. H.

Huxley in his famous Romanes Lecture on " Evolution and Ethics " (1893). Unlike Fiske, Huxley accepted at its face value the Hobbesian interpretation of Darwinism and acknowledged that " men in society are undoubtedly subject to the cosmic process," which includes, of course, the struggle for existence and the elimination of the unfit. But he flatly rejected the common practice of identifying the " fittest " with the " best," pointing out that under certain cosmic conditions the only " fit " organisms would prove to be low ones. Man and nature make altogether different judgments of value. The ethical process, or the production of what man recognizes as truly the " best," is in opposition to the cosmic process. " Social progress means a checking of the cosmic process at every step."

In a companion essay Huxley compared the ethical process to the work of the gardener: the state of the garden is not that of " nature red in tooth and claw," for the horticultural process eliminates struggle by adjusting life conditions to the plant instead of making the plants adjust to nature. Instead of encouraging, horticulture restricts multiplication of the species. Like horticulture, human ethics defies the cosmic process; for both horticulture and ethical behavior circumvent the raw struggle for existence in the interest of some ideal imposed from without upon the processes of nature.

The more advanced a society becomes, the more it eliminates the struggle for existence among its members. To practice natural selection in a society after the fashion of the jungle would weaken, perhaps destroy, the bonds holding it together:

> It strikes me that men who are accustomed to contemplate the active or passive extirpation of the weak, the unfortunate, and the superfluous; who justify that conduct on the ground that it has the sanction of the cosmic process, and is the only way of ensuring the progress of the race; who, if they are consistent, must rank medicine among the black arts and count the physician a mischievous preserver of the unfit; on whose matrimonial undertakings the principles of the stud have the chief influence; whose whole lives, therefore, are an education in the noble art of suppressing natural affection and sympathy, are not likely to have any large stock of these commodities left. But, with-

out them, there is no conscience, nor any self-restraint on the conduct of men, except the calculation of self-interest, the balancing of certain present gratifications against doubtful future pains; and experience tells us how much that is worth.[18]

What is called the struggle for existence in modern society is really a struggle for the means of enjoyment. Only the desperately poor, the pauperized, and the criminal are engaged in a struggle for actual existence; and this struggle among the submerged 5 per cent of society can have no selective action on the whole, because even the members of this class manage to multiply rapidly before they die. The struggle for enjoyment, while it may have a moderate selective action, is in no way analogous either to natural selection or to the artificial selection of the horticulturist. Then the need of mankind is not acquiescence to nature, but " a constant struggle to maintain and improve, in opposition to the State of Nature, the State of Art of an organized polity." [19]

Reminiscent of Fiske's infancy theory were Henry Drummond's popular Lowell Lectures on *The Ascent of Man* (1894). Drummond, a Scottish preacher who had already gained a considerable following with his pseudo-philosophical book, *Natural Law in the Spiritual World* (1883), did not deny the importance of the Struggle for Life, but he looked upon it as the villain of the piece rather than the play itself. A second factor in evolution, equally important, is the Struggle for the Life of Others. The Struggle for Life springs from the requirements of nutrition; reproduction and its resulting emotions and relationships are the foundation of the Struggle for the Life of Others. With Fiske, Drummond found in the family the basis of human sympathy and solidarity, for it is there that the Struggle for the Life of Others begins.

Critical of Huxley's dualism between the cosmic and the ethical, Drummond sought for a natural foundation for moral behavior. His solution was a teleological interpretation of the evolutionary process in which the Struggle for the Life of Others was seen as a Providential device for securing perfection. In this way Drummond killed two birds with one stone: he restored the continuity of natural evolution and morals and saved spiritualism from mechanical interpretations of evolution. "The path of progress and the path of Altruism are one. Evolution is nothing but the Involution of Love, the revelation of Infinite Spirit, the Eternal Life returning to Itself." [20] Drummond recognized the ability to survive as mere fittedness, without reference to ethical values. He acknowledged a certain analogy between the industrial process and evolutionary struggle, and found that industry " is but one or two removes from the purely animal struggle." [21] But with the growing importance of the Struggle for the Life of Others and the advance of technology, the struggle is losing its animal fierceness. While the first few chapters of evolution may be headed the Struggle for Life, the book as a whole is a love story.

Less popular than Drummond's book, but more enduring, was Kropotkin's *Mutual Aid* (1902). This work was originally conceived as an answer to Huxley's " Evolution and Ethics," for Kropotkin had the collectivist's natural hostility to philosophies that neglected to see coöperation as a major factor in evolution. When Kropotkin had been in Northern Asia he had seen an impressive measure of mutual aid among the rodents, birds, deer, and wild cattle of Siberia, which brought forcibly to his mind the absence of a bitter struggle for means of subsistence *among animals belonging to the same species*. Some Darwinists had considered internecine strife a critical factor in evolution, but according to Kropotkin, Darwin had been innocent of this because he had recognized unequivocally the element of coöperation.

Kropotkin backed his thesis with an impressive amount of natural and historical lore, culled from a wide range of literature. From ants, bees, and beetles, through all the mammalia, Kropotkin found sociability and coöperation within the species-unit. Birds, even birds of prey, are sociable, and wolves hunt in packs. Rodents work in common, horses herd together, and most monkeys live in bands. Kropotkin followed this with a survey of mutual aid in man — primi-

tive, barbarian, medieval, and modern. On the lessons of
biology for human life he concluded:

> Happily enough, competition is not the rule either in the animal
> world or in mankind. It is limited among animals to exceptional peri-
> ods, and natural selection finds better fields for its activity. Better
> conditions are created by the *elimination of competition* by means of
> mutual aid and mutual support. . . .
>
> " Don't compete! — competition is always injurious to the species, and
> you have plenty of resources to avoid it! " That is the *tendency* of
> nature, not always realized in full, but always present. That is the
> watchword which comes to us from the bush, the forest, the river, the
> ocean. " Therefore combine — practise mutual aid! That is the surest
> means for giving to each and to all the greatest safety, the best guaran-
> tee of existence and progress, bodily, intellectual, moral." That is
> what Nature teaches us.[22]

III

From other quarters the principle of competition was defended with new subtleties. In the 1890's, although competition was increasingly thrown on the defensive, two popular writers entered the lists on its behalf and once again
attempted to fit competitive ethics into the evolutionary
scheme.

Two new currents in the intellectual atmosphere pro-
voked a change in the tone of evolutionary apologetics: the
growth of social protest evident in the Henry George and
Edward Bellamy movements, the publication of the Fabian
essays, and a growing general familiarity with Marxism; and
in the field of biology the publication of August Weismann's
researches into the inheritance of acquired characteristics.[23]
Weismann had developed what he thought was conclusive
evidence against such inheritance. If he was right — and
most biologists believed he was — the Lamarckian features
of Herbert Spencer's philosophy were no longer tenable;
men could no longer hope to evolve an ideal race by gradual
increments of knowledge and benevolence handed down to
their children; social evolution must be redrawn along
stricter Darwinian lines; if there was to be any progress at all
it must come from a severe reliance upon natural selection.

Benjamin Kidd, an obscure English government clerk, capitalized upon these problems in his *Social Evolution,* which appeared in 1894 and became the rage in the Anglo-American literary world. Kidd attempted to set up a theoretical structure based upon Weismann, which would reconcile the competitive process, natural selection, and the trend toward legislative reform initiated by the new protest. His theory started with the familiar dogma that progress results from selection and that selection inevitably involves competition.[24] Therefore the central aim of a progressive civilization must be to maintain competition.

For the great masses of men, however, for the underdogs everywhere, the incentives to maintain competition grow slighter and slighter, Kidd realized. Hence the swelling cry of social protest.

[Man's] interests as an individual have, in fact, become further subordinated to those of a social organism, with interests immensely wider and a life indefinitely longer than his own. How is the possession of reason ever to be rendered compatible with the will to submit to conditions of existence so onerous, requiring the effective and continual subordination of the individual's welfare to the progress of a development in which he can have no personal interest whatever?[25]

Why should the red Indian or the New Zealand Maori, undergoing extermination before the advance of more progressive peoples, have an interest in progress? Or, more important for western civilization and its future, what rational sanction can there be for the "great masses of people, the so-called lower class," to submit to the personal trials and tortures incident to social progress by way of the competitive system? They are already becoming more and more aware that their individual rational interest is clearly to abolish competition, to suspend rivalry, to establish socialism, to regulate population and keep it "proportional to the means of comfortable existence for all."

This antagonism between the rational interest of the mass-individual and the continued progress of the social organism, Kidd argued, cannot be reconciled by reason. But let philosophy abandon its attempt to find a rational sanction for human conduct — then the problem is seen in a new light. At the same time the social function of religion is made crystal clear.

One common characteristic underlies all conceptions of religion: they reveal " man in some way in conflict with his own reason." The universal instinctive religious impulse serves this indispensable social function: it provides a supernatural, nonrational sanction for progress. All kinds of religious systems are " associated with conduct, having a social significance; and everywhere the ultimate sanction which they provide for the conduct they prescribed is a super-rational one." Religion as a social institution has survived because it performs an essential service to the race: it impels man to act in a socially responsible way. Such an impulse is absent from all merely rational ways of thought.[26]

For the role of altruism in human affairs, Kidd had a defense notably different from Spencer's. There is no rational sanction for altruism; its sanction is superrational, and runs counter to individual self-interest. No wonder that it is so often found in close association with the religious impulse. The altruistic impulse should be heeded, and is being heeded, for there is a growing tendency to strengthen and equip the lower and weaker against the higher and wealthier classes of the community. This is the best possible answer to the threat of socialism. Socialism, abandoning competition, would result in degeneracy and inundation by more vigorous societies. The effect of charities, and of the general trend toward strengthening the masses to compete by means of social legislation is to stimulate competitive tension. Thus the social efficiency of western society is increased. All future progressive legislation must lift the masses into this energetic competition. As state interference widens, mankind will paradoxically move further and further away from socialism. The state will never go so far as to manage industry or confiscate private property.[27] From all this progressive movement will come a " new democracy " higher than anything yet attained in the history of the race.

It was a peculiar mixture of obscurantism, reformism,

Christianity, and social Darwinism that Kidd offered his thousands of readers. Among religious folk who wanted a rational foundation for their beliefs, among social Darwinists of the older laissez-faire stripe, orthodox Spencerians, trained philosophers and sociologists, and rationalists of all kinds, Kidd's doctrines were anathema. But this hostility did not prevent his having tremendous popular appeal. " His reputation," complained an eminent American sociologist, " seems to me one of the most humiliating freaks of book-readers' opinion that has occurred in the generation that put Mrs. Humphrey Ward on a pedestal and is now incoherently Trilby-mad." [28] A more patient explanation was offered by John A. Hobson in the *American Journal of Sociology:*

> There has been a rapidly growing feeling among large numbers of those who still cleave to the orthodox churches, that the intellectual foundations of religion have slipped away. They are not rationalists, most of them have never seriously examined the rational basis of their creed, but the disturbing influences of rational criticism have reached them in the shape of this vague uneasy feeling. Now these people, morally weak because they have relied upon dogmatic supports of conduct, are ready to grasp eagerly at a theory which will save their religious systems in a manner which seems consistent with the maintenance of modern culture.[29]

A mixed reaction was expressed by Theodore Roosevelt in the *North American Review.* He approved of Kidd's assertion that social progress continues to rest upon biological laws; of his attack on socialism as retrogressive; of his conclusion that the state should equalize the chances of competition but not abolish it; of his emphasis on efficiency as a criterion of society, and on character as opposed to intellect. He felt, however, that Kidd overstressed the necessity of competition and understressed the tendency of the unfit, even without organized social aid, to survive and grow more fit rather than suffer elimination. He also argued that Kidd exaggerated the sufferings of the masses; in a progressive community four-fifth or nine-tenths of the people are happy and therefore do have a rational sanction for contributing to progress. Moreover, Kidd valued all religions alike, whereas Christian-

ity is far superior to others in teaching the subordination of
the individual to the interests of mankind. Finally, Roosevelt was not happy with Kidd's view of religion, which he
found equivalent to calling it " a succession of lies necessary
to make the world go forward." [30]

Four years later, William H. Mallock, an English hack
writer who was well known in this country for his books and
magazine articles, brought out a volume entitled *Aristocracy
and Evolution,* in which he proposed to throw overboard the
whole of Kidd's system, along with the rest of the prevailing
evolutionary social theory, and return to pure individualism.

Mallock's intention was to establish the rights and the
social functions of the wealthier classes, which he felt had
been inadequately understood in the evolutionary philosophy
of Spencer and Kidd. The great fault of current sociology
was that it spoke grossly of "mankind" or "the race" or
"the nation," without refining these terms into classes and
individuals. With all their talk about the evolutionary progression of the whole mass of society, Spencer and Kidd were
particularly guilty of disparaging the great man and losing
sight of his contributions and achievements. They fallaci-
ously belittled the stature of great leaders by attributing their
deeds to the whole of society and its inherited skills and accomplishments; by the same logic the great masses of men
could also be shorn of credit for their petty performances.

The great man, in Mallock's scheme, was certainly not to
be identified with the physically fittest survivor in the struggle for existence. All you could say for the physically fittest
survivor was that he manages to live; and while this does
undoubtedly contribute to the progress of the race, it is
slow and unspectacular. The great man, on the other
hand, galvanizes society by acquiring unique knowledge
or skill and imposing it on the mass. The physically
fittest promotes progress by living while others die; the
great man promotes progress by helping others to live.
The struggle of ordinary workers to find employment is a
social equivalent of the struggle for existence; it contributes
but little to progress, for the greatest forward steps in the de-

velopment of man have been accomplished without any improvement in the breed of its laborers. The industrial struggle that really promotes progress is the battle among leaders, among employers. When one of two competing employers succeeds in conquering the other, the working men of the vanquished are absorbed in the employ of the victor, and lose nothing; but the fruits of the successful leader's skill are bequeathed to the community. It is, then, not the brute struggle for existence but the war for domination among the well-to-do that results in social progress.

Domination by the fittest is of the greatest benefit to society as a whole. In order to facilitate the process the great man must be impelled by strong motives and granted the instruments of domination. Fundamentally this is an economic problem. The great man can exert his influence by one of two economic means — the slave system and the capitalistic wage system, the one a system of compulsion, the other of voluntary inducement. Socialists, who desire to abolish the wage system, can do so only by founding a slave system. They could not eliminate the struggle for domination; they could only enclose it in their cumbrous and wasteful order. To progress, a social system must retain competition between the directors of labor, the contest for industrial domination. No matter what happens to society, the domination of the fittest great men — capitalistic competition — must be ensured. Such men are the true producers. The fundamental condition of social progress is that these leaders be obeyed by the masses. In politics, as in industry, the forms of democracy are hollow; for while executive agencies are designed to execute the will of the many, the opinions of the many are formed by the few, who manipulate them.[31]

IV

A reader who had followed with equal devotion and equal credulity the suggestions of all these writers might have felt his confusion growing instead of being resolved. Yet amid all this confusion there was a decided trend, most evident when one considers what was agreed upon by Fiske, Drummond, and Kropotkin. They all endorsed solidarism; they saw the group (the species, family, tribe, class, or nation) as the unit of survival, and minimized or overlooked entirely the individual aspect of competition. It was precisely this which Mallock, an arch-individualist, found objectionable in the current trend of evolutionary thought. Fiske, Drummond and Kropotkin not only agreed that social solidarity is a basic fact in evolution, but believed further that solidarity is a thoroughly natural phenomenon, a logical outgrowth of natural evolution.[32] In this respect they differed from Huxley, who shared their concern about the effect of the struggle-for-existence philosophy upon " the social bond." But Huxley, finding no basis in the " cosmic process " for the " ethical process," was obliged to tear the two asunder and establish a dualism of facts and values which aroused a great deal of criticism. Even Kidd's devotion to competition in the abstract was qualified by his acceptance of social legislation in the interest of group efficiency.

The transition to solidarism, which was part of a larger reconstruction in American thought, became apparent in the 1890's — the period that saw the publication of Drummond and Kidd, of Huxley's essay, and, in preliminary form, of *Mutual Aid*. Rising with solidarism were other streams of criticism. In the realm of philosophy, the new spirit was marked by the ascendancy of the pragmatic movement, especially significant because it rejected the cold determinism of Spencer's philosophy and constructed a new psychology, in part out of Darwinian materials. As social dissent became more vociferous, there arose a new concern with conscious social control. Inspired by events in the political and industrial arena, social science also reassessed its aims and methods. Earlier conceptions of the social significance of Darwinism were undergoing profound changes.

Chapter Six

The Dissenters

We may go far beyond Mr. Spencer's limits and yet stop a great
way this side of socialism.

WASHINGTON GLADDEN

The sincere and candid reformer can no longer consider the national Promise as destined to automatic fulfillment. The reformers . . . proclaim their conviction of an indubitable and a beneficent national future. But they do not and cannot believe that this
future will take care of itself. As reformers they are bound to
assert that the national body requires for the time being a good
deal of medical attendance, and many of them anticipate that even
after the doctors have discontinued their daily visits the patient
will still need the supervision of a sanitary specialist.

HERBERT CROLY

From the disorders and discontents that plagued America
in the seventies, eighties, and nineties, there arose a stream
of dissenting opinion on the merits of the free competitive
order. Two panics followed by long and harrowing depres-
sions racked the economic life of the nation in the first and
last of these decades; and in the intervening one, hardly a
period of uninterrupted prosperity, labor uprisings of un-
precedented scope and violence took place. The growth of
the Knights of Labor and the strikes of the eighties, climaxed
by the eight-hour movement and the Haymarket affair, gave
to labor strife a central place in public attention. In the
depression of the nineties, agricultural protest combined with
labor unrest to create the national political upheaval of 1896.

Outside the immediate ranks of labor, an articulate source
of reform sentiment in urban communities was the social-
gospel movement. Many Protestant clergymen now criticized industrialism as their predecessors had criticized slavery,
and their protest gave to the dissent of the post-bellum period
a strong Christian flavor.

The clergy of the cities had direct experience with industrial evils. They saw the living conditions of the workingmen, their slums, their pitiful wages, their unemployment, the enforced labor of their wives and daughters. Many ministers were troubled because the churches were out of touch with the working class, and sensed the unreality of talk about moral reform and Christian conduct in such an oppressive and brutalizing environment. They were not only shocked but alarmed by the industrial scene. Although they sympathized with trade unions, especially as defensive organizations, they were troubled by the ugly potential of industrial violence. They were learning about the doctrines and methods of European socialism, and, at the outset at least, feared their spread in the United States. What they sought, therefore, was a compromise between the harsh individualism of the competitive order and the possible dangers of socialism. Although agrarian discontents played a prominent part in national and state politics, the clergy focused their attention almost exclusively upon the problems of labor. There lay the menace; there lay the promise.[1]

Most social gospel leaders worked in this urban setting. The most famous and the most active of them was the prolific Washington Gladden (1836–1918), a preacher in several cities and for a time a writer on the editorial staff of the *Independent.* Among Gladden's contemporaries who shared his moderate reformism were Lyman Abbott, one of the most influential clergymen of the age; the Rev. A. J. F. Behrends, who hoped to persuade Christians to forestall the menace of socialism by anticipating its more acceptable proposals; and Francis Greenwood Peabody, who taught Christian ethics at Harvard. Other advocates of the social gospel were closer to socialism. William Dwight Porter Bliss (1856–1926) of Boston organized a Protestant Episcopal reform group, the Church Association for the Advancement of the Interests of Labor (CAIL), and published a radical paper, the *Dawn*, which supported sundry left-wing movements. George Herron (1862–1925), a famous platform speaker and professor of Applied Christianity at Iowa College who joined the So-

cialist Party in 1889, was a leading propagandist of the movement. Walter Rauschenbusch (1861–1918), another convert
to socialism, exerted through his writings a profound influence on Christian social thinking in the Progressive period.

The greatest literary successes of the movement were produced by midwesterners. Josiah Strong's discussion of national problems, *Our Country*, was a best seller in the 1880's.
A Kansas minister, Charles M. Sheldon, wrote a crudely
novelized tract, *In His Steps* (1896), describing the social
experiences of a small-town congregation that patterned its
conduct on the precepts of Jesus; the volume sold about
23,000,000 copies in English between the day of its publication and 1925.[2]

The movements inspired by Henry George and Edward
Bellamy were of one piece with the social gospel. Both of
these men, products of pious home environments, were intensely religious; their writings were filled with a moral protest thoroughly familiar to readers of social-gospel literature.
That the social gospel and the followers of George and Bellamy shared a common outlook was shown by the adherence
of many socially-minded clergymen to both the Nationalist
and single-tax movements. On another front the social gospel was linked to those academic economists who had begun
to criticize individualism; such progressive economists as
John R. Commons, Edward Bemis, and Richard T. Ely
formed a bridge between churchmen and other professional
economists. At one time over sixty clergymen were listed as
members of the American Economic Association.[3]

The social-gospel movement arose during the years when
evolution was making converts among the progressive clergy,
and since ministers who were liberal in social outlook were
almost invariably liberal in theology also, the social theory
of the movement was deeply affected by the impact of naturalism upon social thought. The growing secularization of
thought hastened the trend among clergymen to turn from
the abstractions of theology to social questions. The liberali-
zation of theology broke down the insularity of religion.
Social-gospel leaders were also inspired by the vistas of de-

velopment opened both forward and backward in time by the evolutionary perspective; their belief in an inevitable progress toward a better order on earth — the Kingdom of God — was fortified by the evolutionary dogma. Wrote Walter Rauschenbusch:

Translate the evolutionary theories into religious faith, and you have the doctrine of the Kingdom of God. This combination with scientific evolutionary thought has freed the kingdom ideal of its catastrophic setting and its background of demonism, and so adapted it to the climate of the modern world.[4]

Spencer's organic interpretation of society also appealed to the progressive clergy, although they usually put it to uses of which he would have sternly disapproved. To them the social-organism concept meant that the salvation of the single individual had lost its meaning, and that in the future men would speak with Washington Gladden of " social salvation." It also implied a harmony of interests between classes which served as a framework for their appeals against class conflict and for extended state intervention.[5] Lyman Abbott, however, thought that the social-organism idea provided an argument for slow and gradual reform.[6] No longer under the influence of the theological concept of the total depravity of human nature, some social-gospel writers also accepted the idea that the social order should be transformed by changing the character of individuals — a conception in which they were close to Spencer and other conservatives.

In one critical respect the pioneers of the social gospel departed from prevailing social uses of evolution: they detested and feared the free competitive order and all its works. However profoundly influenced by individualism, however timorous about socialism, they were in general agreement on the need to modify the free workings of competition, to abandon Manchesterian economics and the social fatalism of the Spencerians. " Christianity," wrote the Rev. A. J. F. Behrends, " cannot grant the adequacy of the ' laissez-faire ' philosophy, cannot admit that the perfect and permanent social state is the product of natural law and of an unrestricted competition." [7] Citing Emile de Laveleye, a Belgian expositor of socialism, as having said that followers of Darwin and advocates of a natural-law political economy " are the real and only logical adversaries at once of Christianity and of socialism," Behrends continued:

> Our contention is not against Darwinism as a philosophy of unconscious and irresponsible existence; it may be true in purely biological science; but the gifts of reason and of conscience, the powers of self-consciousness and of self-determination, make man more than an animal or a plant, and so invest him with the power to modify and control the law of natural selection and to mitigate the fierceness of the struggle for existence. . . .
>
> It is time that the poor and oppressed should understand that their deliverance will never come from the political economy which allies itself with the school of Haeckel and Darwin. It knows nothing of the duty of mercy, it recognizes only the right of the fittest to survive.[8]

Of like mind was Washington Gladden, who often asserted his opposition to Spencer and all the glorifiers of selective competition. He warned that the weaker classes would unite to attack a competitive system in which they were threatened with annihilation, and that huge warring combinations of capital and labor would be the natural consequence of accepting the law of strife as a norm for industrial society.[9] He urged an " industrial partnership " between employers and employees as an alternative to disaster. If binding natural laws were conceived to govern economic behavior, it would be futile to urge employers to obey the promptings of heir Christian conscience and deal more generously with heir men.[10] Expressing a desire for the growth of trade unions to balance large industrial combinations, he hoped that arbitration would supersede strife as the means of settlement. The principle of competition, the survival of the fittest, is the law of plants and brutes and brutish men, but it is not the highest law of civilized society. The higher principle of good will, of mutual help, begins to operate in the social order, and the struggle for existence disappears with the progress of the race.[11]

More fervid in his attacks upon self-interest and strife as bases of social organization, George Herron pilloried Sumner and Spencer for their appeal to self-interest.[12] To assume placidly that competition is the law of life and development was, to Herron, " the fatal mistake of the social and economic sciences." Cain, he declared, had been " the author of the competitive theory." [13]

The most common counterbalance to the competitive principle, in the minds of these leaders, was the principle of Christian ethics and the dicta of the Christian conscience. As Herron expressed it, "the Sermon on the Mount is the science of society." [14] Yet they also welcomed the efforts of men like Fiske and Drummond to find a foundation in the natural processes of evolution for their belief in the limitations of competition as a rule of human life.[15]

As the social gospel developed, it became increasingly cordial to municipal socialism or public regulation of basic industries; this could be seen in the writings of many who had the conventional objections to socialism. To the growing solidaristic trend in American thought the social gospel contributed heavily, for its lectures were heard by thousands, its books read by hundreds of thousands, and incalculable numbers joined its organizations or attended its earnest conferences. A current of criticism frequently neglected and underrated by historians of American social literature, it supplied several religious bodies with a lasting reform orientation, and paved the way for all socially-minded Protestant movements of a later day. Not the least of its accomplishments was to break ground for the Progressive era.

II

The two most outstanding spokesmen of urban discontent, Henry George and Edward Bellamy, felt the necessity of refuting the conservative arguments of evolutionary sociology. Henry George differed from other dissenting ideologists in his acceptance of competition as the necessary way of economic life.[16] Like most other dissenters, however, he found himself compelled to grapple with the fatalism of the evolutionary sociology. If the single tax on land values were to be accepted as the open-sesame to a new world of progress and plenty, George felt he must first refute both the Malthusian explanation of misery and the Spencerian argument against rapid progress. Thus the second book of his great work *Progress and Poverty* was devoted to disproving Malthus, who, George believed, still had a firm grip upon many economic thinkers. George pointed to the coexistence of want with the highest productive powers as evidence that the Malthusian pressure of population upon subsistence had not begun to operate.

He concluded:

I assert that the injustice of society, not the niggardliness of nature, is the cause of the want and misery which the current theory attributes to overpopulation. I assert that the new mouths which an increasing population calls into existence require no more food than the old ones, while the hands they bring with them can in the natural order of things produce more.[17]

In the last section of *Progress and Poverty*, George confronted the prevailing evolutionary conservatism. The actual outcome of the doctrine of progress as implemented by the Darwinian struggle for existence, he wrote, " is in a sort of hopeful fatalism, of which current literature is full."

In this view, progress is the result of forces which work slowly, steadily and remorselessly, for the elevation of man. War, slavery, tyranny, superstition, famine, and pestilence, the want and misery which fester in modern civilization, are the impelling causes which drive men on by eliminating poorer types and extending the higher; and hereditary transmission is the power by which advances are fixed, and past advances made the footing for new advances. The individual is the result of changes thus impressed upon and perpetuated through a long series of individuals, and the social organization takes its form from the individuals of which it is composed

Herbert Spencer had said in *The Study of Sociology* that this theory of society was " radical to a degree beyond anything which current radicalism conceives," since it anticipates a change in human nature itself; but, said George, it is also onservative beyond anything conceived by current conservatism because " it holds that no change can avail, save these slow changes in men's natures." This theory, which represents the prevailing view of civilization,[18] accounts neither for the failure of some peoples to progress (a problem Bage-hot tried to solve) nor for the failure of others to maintain a level of civilization once achieved. History suggests that civilizations rise and fall in a wavelike rhythm. It is pos-sible that each national or race life has a stock of energy which it expends; as the energy is dissipated the nation de-clines. But George thought that he had a better explana-tion: " That the obstacles which finally bring progress to a halt are raised by the course of progress; that what has de-stroyed all previous civilizations has been the conditions produced by the growth of civilization itself." [19]   The prin-cipal conditions of social progress are association and equal-ity, and society is now threatened by the division and inequality it breeds. The seeds of the destruction of the existing order could be found in its own poverty; in its squalid cities were already breeding the barbarian hordes which might overwhelm it. Civilization must either prepare itself for a new forward leap or plunge downward into a new barbarism.[20]

When he wrote *Progress and Poverty,* George was familiar with the arguments against private land ownership expressed by Spencer in *Social Statics,* and he treasured the hope that he might be able to bring behind the force of his movement the authority of the great philosopher. Spencer's failure to acknowledge the copy of *Progress and Poverty* sent to him might have served as a warning of the disappointment to come. During the course of his trip to the British Isles in 1882, George met Spencer at the home of H. M. Hyndman, and their conversation turned at once toward the agitation of the Irish Land League, which had commanded George's sympathy. Spencer wasted no time in telling George that the imprisoned Land League agitators had got what they de-served, whereupon George's opinion of the philosopher changed completely. Ten years later, after Spencer had per-mitted the issuing of a revised and abridged edition of *Social Statics* from which the attacks on land ownership were re-moved, George settled the score by publishing a lengthy attack on Spencer under the title *A Perplexed Philosopher.*

While the volume was primarily a review of the allegedly discreditable motives of Spencer's retraction, George also assailed the callousness of the Spencerian political philosophy as it was expressed in *The Man Versus the State.* In these essays, he declared, " Mr. Spencer is like one who might insist that each should swim for himself in crossing a river, ignoring the fact that some had been artificially provided with corks and others artificially loaded with lead." [21]

The Nationalist movement that sprang up after the publication of Bellamy's *Looking Backward* in 1888 centered its fire not upon the land question but upon the fundamental principle of the competitive system and upon the institution of private property itself. When Julian West, the hero of *Looking Backward,* awoke in the year 2000 to find himself living in Bellamy's mechanical Utopia, one of his first reactions was to comment, " Human nature itself must have changed very much," to which his host, Dr. Leete, replied, " Not at all, but the conditions of human life have changed, and with them the motives of human action." [22] As the wonders of the coöperative order unfolded, it became clear to Julian West that this change of conditions centered about the abolition of strife. " Selfishness was their only science," complained Dr. Leete of the men of the nineteenth century, " and in industrial production selfishness is suicide. Competition, which is the instinct of selfishness, is another word for dissipation of energy, while combination is the secret of efficient production." [23]

The " Declaration of Principles " of Bellamy's Nationalist movement (which derived its name from his proposal to nationalize industry) began:

The principle of the Brotherhood of Humanity is one of the eternal truths that govern the world's progress on lines which distinguish human nature from brute nature.

The principle of competition is simply the application of the brutal law of the survival of the strongest and most cunning.

Therefore, so long as competition continues to be the ruling factor in our industrial system, the highest development of the individual cannot be reached, the loftiest aims of humanity cannot be realized.[24]

In an address before a Boston audience Bellamy declared that " the final plea for any form of brutality in these days is that it tends to the survival of the fittest; and very properly this plea has been advanced in favor of the system which is the sum of all brutalities." If the richest were in fact the best, he continued, there would have been no social question, and disparities of condition would have been willingly endured; but the competitive system apparently causes the unfittest to survive, not in the sense that the rich are worse than the poor, but that the system encourages the worst in the character of all classes.[25]

Similar attacks upon competition or individualism were common in nationalist literature.[26] When Lester Ward published his article on " The Psychological Basis of Social Economics " in which he elaborated the distinction between animal and human economics, Bellamy wrote him a note of warm approval suggesting that some way be found to give the article general circulation. Subsequently he republished the greater part of it in his second Nationalist magazine, the *New Nation*. " It will bear study," he advised his readers, " as furnishing the best of ammunition for replying to the ' survival of the fittest ' argument against nationalism." [27]

American socialist writers tried persistently to show that evolutionary biology does not provide a justification for competitive individualism. Laurence Gronlund, who was at one time close to the Nationalist movement and later an officer of the Socialist Labor Party, took great pains to distinguish between the healthy " emulation " that would go on in a coöperative commonwealth and the unhealthy competition of capitalism. In his work, *The Coöperative Commonwealth* (1884), he used Spencer's idea of the social organism to refute Spencer's individualism. The organic character of social life, he argued, demands increasing centralization and management.[28] Now all but forgotten, Gronlund's writings were once widely read by intellectuals interested in socialism, who seem to have found satisfaction in his occasional religious phraseology, his moderate tone, his air of theoretical authority. From him the social-gospel prophets drew many of their ideas. Gronlund's book *Our Destiny* (1890), which was published in an abridged version in Bellamy's *Nationalist*, was an assault upon the ethics of competition as conceived by Spencer and his followers. In language not unlike that of Ward, whose *Dynamic Sociology* he had read, Gronlund insisted that conscious evolution would be a far different thing from the unmodified natural evolution of the past, and that human intervention must play an increasingly important role in development. Having also read Marx, Gronlund asserted that the rise of trusts was paving the way for socialism, and that the continuing " trustification " of industry was a proof of the superiority of combination over competition. While Gronlund had always criticized the " fatalistic " aspect of Spencer's social theory, he urged his readers to believe that combination was the " inevitable " next step in social evolution, leaving them a choice between monopolized capitalism and a collectivized social order.[29]

III

Orthodox Marxian socialists in the early years of the twentieth century felt quite at home in Darwinian surroundings. Karl Marx himself, with his belief in universal " dialectical " principles, had been as much a monist as Comte or Spencer. Reading *The Origin of Species* in 1860, he reported to Friedrich Engels, and later declared to Ferdinand Lassalle, that " Darwin's book is very important and serves me as a basis in natural science for the class struggle in history." [30] On the shelves of the socialist bookstores in Germany the works of Darwin and Marx stood side by side. American socialist intellectuals were quick to adopt the latest developments in the realm of scientific knowledge, and the little green-bound volumes that came pouring forth from the Kerr presses in Chicago were frequently adorned with knowing citations from Darwin, Huxley, Spencer, and Haeckel. Arthur M. Lewis' lectures at the Garrick Theater on the relations of science and revolution were extraordinarily well attended; in book form, under the title *Evolution, Social and Organic* (1908), they went through three editions and had the largest

advance sale of any native socialist publication.[31]   The concern with the problem among socialist intellectuals was reflected in the early volumes of the *International Socialist Review,* whose contents suggest that socialists considered " scientific " individualism a sufficiently live doctrine to be worth
refuting.   One of them referred to natural selection as " the
last remaining bulwark of the fortress of individualism." [32]

As Marx had found in the struggle for existence a " basis "
for the class struggle, American socialists found even in the
writings of Spencer aid and comfort for their cause.   They
approved of the idea of the social organism, and, like Gronlund, they turned it to their own uses: they lauded Spencer's
attack upon the great-man theory of history; they approved
of his agnosticism; they were indebted to him for having
helped to persuade the world that society was in flux along
with the rest of organic life.[33]   His individualism they naturally considered inconsistent with the body of his scientific
teachings; and they sought to drive a deep wedge between
the evolutionary Spencer who conceived of the social organism and the individualistic Spencer of *The Man Versus the
State.*[34]

Post-Darwinian trends in biological theory were hailed by
socialists as definitive proof of the validity of their approach.
Ward and Spencer, relying respectively upon education and
gradual character development as the media of social improvement, were discouraged at the abandonment of Lamarckian use-inheritance; but the socialists, hoping for reconstruction of the economic environment, found Weismann's theory more congenial.   Wrote Lewis:

> If it were true that the terrible results of the degrading conditions
> forced upon the dwellers in the slums were transmitted to their chil-
> dren by heredity, until in a few generations they became fixed charac-
> ters, the hope of Socialists for a regenerated society would be much
> more difficult to realize.   In that case these unfortunate creatures would
> continue to act in the same way for several generations, no matter how
> their environment had been transformed by the corporate actions of
> society.   This much at any rate, Weismann has done for us, he has
> scientifically destroyed that lie.[35]

Even more to their liking was the *Mutationstheorie* of Hugo
deVries. DeVries, a Dutch biologist, had contributed to the
solution of one of the difficulties in natural-selection theory
by pointing to the role of " sports " or mutations, sudden
and drastic variants in individual organisms, in the process
of adaptation. Biologists had been introduced by DeVries'
theory to a new view of the evolutionary process as catastrophic and abrupt — a strong contrast to the slow, legato,
and minuscule variations of Darwin's evolution. In social
theory, Darwin's view had shored up the argument for " the
inevitability of gradualness " which had been so prominent in
the conservatism of Spencer and Sumner. " For a half a century," explained Lewis, " this argument of slow evolution has
done valiant service as an antidote for Socialism, and the
present ruling class would like to retain it forever." The
mutation theory made it clear, however, that nature's method
is to alternate periods of gradual evolution with sudden
" revolutionary " spurts. The social equivalent of this was
the sudden and drastic reconstruction of the economic basis
of society proposed by Marxists.[36] Lewis also made good use
of Kropotkin's *Mutual Aid* and the writings of Lester Ward.[37]

While the socialists were adept at seizing and synthesizing
the standard criticisms of nineteenth-century " evolutionary " sociology, they had little that was new or original to say.
Sound as the socialist criticisms may have been, they were
nevertheless stereotypes, cast in the mold of the same nineteenth-century monism that afflicted both Marx and Spencer.
Only when biology seemed to agree with their social preconceptions were they ready to build a sociology upon it.
They were willing to use the struggle for existence to validate
the class struggle, but not individualistic competition. They
objected to Darwinism as a conservative rationale, but they
saw nothing wrong with the conception of a biologically
centered social theory if it could be fastened to their own
system. The most independent socialist work in this respect,
William English Walling's *The Larger Aspects of Socialism*,
appeared in 1913. With his comrades, Walling rejected the

conservative conclusions of biosocial speculation, but the form of his argument was different: he relied upon the anthropocentric humanism of James and Dewey and tried to merge the philosophies of socialism and pragmatism. His object was to set the new experimental approach against the absolutism of the nineteenth-century nature philosophers and all the arguments that had grown out of their monistic assumption.

While other socialists had merely argued that current conservative applications of biology to sociology were inept, Walling in a more sweeping fashion attacked the common tendency to base social theory on biology. Not only did he object to the " optimistic fatalism " of Spencer and the argument for competition from natural selection, but he also rejected the social-organism analogy. He felt that it encouraged an emphasis on the race or the state at the expense of the individual; this was inconsistent with the humanistic aims of a true socialism. Instead he urged that the processes of social evolution be conceived as *qualitatively different;* and that the emphasis be placed upon changes in the environment wrought by the creative agency of man rather than upon the more passive process of adaptation to an environment conceived as fixed and final. He concluded:

We are chiefly interested not in the " origin of species " in nature, but in the destiny of species under man, not in the " creative evolution " of nature, but in the infinitely more creative evolution of man. Our affair is not with the evolution of life and its adaptation to the natural environment, but with the evolution of man, and the adaptation of life to his purposes. And even the control of the life around us matters less than the control of our own lives, and the control of our physiological evolution less than that of our psychological evolution and of social progress.[38]

IV

Of course the reform groups never had the satisfaction of seeing their plans put into effect, but their efforts to contest the intellectual premises of unrestrained individualism did meet with some success. If no Utopia was in the making, at least there was a shift away from the free competitive order. The material basis of the Spencer-Sumner ideology was being transformed, and the lines of social argument were pushing on. It was not so much that the old arguments for individualism had been answered to general satisfaction; they had been swept away by a ground swell of popular feeling deeper than any of the subtleties of social theorists. As new contestants came upon the scene, the focus of the debate changed.

Populists, Bryanites, muckrakers, progressives, followers of the New Freedom, men and movements transcending by far the influence of socialists, single taxers, and benevolent preachers, took up the causes of reform. The relatively untrammeled capitalism of the nineteenth century was beginning to change into the welfare capitalism of the twentieth; the frustrations of the middle class and the needs of the poor were accelerating the change.[39] Men sensed that a different order was slowly arising. Although they could seldom describe it, they expressed it variously in their slogans and titles: they spoke of the New Nationalism, the Square Deal, the New Freedom, the New Competition, the New Democracy — and, in time, of the New Deal.

Previous reform and protest movements had been disjointed and uncoördinated uprisings of workers and farmers; now the middle class was drawn into the fray. The middle-class citizen, as producer and consumer, was beginning to feel the growth of monopoly and to fear that he would be ground between large combinations of capital and labor. As the middle class worried about maintaining its status and its standard of living, the figure of the great capitalist entrepreneur, hitherto heroic, lost much of his glamour. He was condemned as an exploiter of labor and an extorter from the consumer, pilloried as an unfair competitor, and exposed as a corrupter of political life. In a society of great collective aggregates, the traditional emphasis upon the exploits of the individual lost much of its appeal. The old problem of defending competition from critics on the left now paled as people were forced to face " the curse of bigness," the more imminent threat to competition from the offspring of compe-

tition itself. " Our industry," complained Henry Demarest
Lloyd in the first major document of the new protest,

. . . is a fight of every man for himself. The prize we give the fittest
is monopoly of the necessaries of life, and we leave these winners of
the powers of life and death to wield them over us by the same "self-
interest " with which they took them from us. . . . " There is no hope
for any of us, but the weakest must go first," is the golden rule of business. There is no other field of human associations in which any such
rule of action is allowed. The man who should apply in his family or
in his citizenship this "survival of the fittest " theory as it is practically
professed and operated in business would be a monster, and would be
speedily made extinct.[40]

It was laissez faire as a policy that was most completely
discredited. While the old, simple apotheosis of competition had faded, few had ceased altogether to believe in it.
One of the primary aims, indeed, of the middle-class revolt
was to restore so far as possible the pristine conditions of
competitive business. But it had become evident, as Lester
Ward had long ago predicted,[41] that even if the supposed
benefits of competition were to be retained, some form of
government regulation was needed to restrain monopoly.
Declared Woodrow Wilson, echoing the complaint of the
little man:

American industry is not free, as once it was free. . . . The man with
only a little capital is finding it harder to get into the field, more and
more impossible to compete with the big fellow. Why? Because the laws
of this country do not prevent the strong from crushing the weak.[42]

Thus the small entrepreneur and his sympathizers, trying
to change the laws, supported a variety of measures between
1904 and 1914 designed to put teeth in the Sherman Act
and otherwise restrain the process of combination. Walter
Weyl explained the changing perspective of individualism:

The little individualist, recognizing his individual impotence, realiz-
ing that he did not possess within himself even the basis of a moral judgment against his big brother, began to change his point of view. He
no longer hoped to right all things by his individual efforts. He turned
to the law, to the government, to the state.[43]

The necessity for state interference was accepted both by those who adhered to the Wilson-Brandeis-LaFollette view that competition is inherently desirable and the Roosevelt-Croly-Van Hise thesis that concentration is inevitable. As Brandeis formulated the governmental problem in 1912:

> . . . the right of competition must be limited in order to preserve it. For excesses of competition lead to monopoly, as excesses of liberty lead to absolutism. . . .
> The issue therefore is: Regulated competition *versus* regulated monopoly.[44]

As serious attempts to alter the business structure through legislation increased, there came a flood of laws to relieve the working class. Intellectuals, humanitarians, and social workers threw themselves on the side of labor, and drew support from a middle class which had no desire to see industrial oppression bring collectivism from the left. In increasing numbers state legislatures adopted laws limiting child and female labor, workmen's compensation laws, and similar measures of reform.[45] Sympathy for union activity grew stronger among intellectuals. In one of his later Massachusetts decisions the austere Oliver Wendell Holmes turned the tables upon the evolutionists by declaring the strike " a lawful instrument in the universal struggle of life." While he believed that labor organizations secured economic gains at the expense of unorganized workers, he considered it an unwarranted conclusion that such activity could be declared illegal. The masses as well as the classes must be judged impartially through the arbitrament of the universal struggle.[46]

The state was conceived by all reformers to be an indispensable instrument of the new reconstruction. In his commentary on *The Promise of American Life* (1909) , a major expression of Progressive thinking, Herbert Croly made a fervent plea for the abandonment of the traditional American " mixture of optimism, fatalism, and conservatism " in favor of a more positive attempt to realize the national promise. He urged that Americans learn to think in terms of purpose rather than destiny, and, without fear of the centralizing powers of government, to realize their purpose through a national policy. The positive tone of the new state capitalism was also expressed by his colleague, Walter Lippmann:

We can no longer treat life as something that has trickled down to us. We have to deal with it deliberately, devise its social organization, alter its tools, formulate its method, educate and control it. In endless ways we put intention where custom has reigned. We break up routines, make decisions, choose our ends, select means.[47]

The managed society which Ward had anticipated and which Sumner had so stoutly opposed was becoming a reality. Little wonder that Spencer had been depressed in his declining years, and well for him that he never lived to see state intervention come of age. Despite the interruption of the twenties, the trend toward social cohesion kept growing,[48] and the sons of the generation that applauded Spencer witnessed the creation of a state machinery as great as any that could have appeared in the Victorian individualist's worst nightmares.[49] Whatever the human potentialities of this apparatus, for good or evil, the ideals of a cohesive and centralized society became increasingly triumphant over those of the heyday of individualism. While individualism had by no means disappeared, it was increasingly on the defensive. As a New Deal leader put it thirty years after Spencer's death:

The religious keynote, the economic keynote, the scientific keynote of the new age must be the overwhelming realization that mankind has such mental and spiritual powers and such control over nature that the doctrine of the struggle for existence is definitely outmoded and replaced by the higher law of coöperation.[50]

Chapter Seven

The Current of Pragmatism

Long after "pragmatism" in any sense save as an application of his *Weltanschauung* shall have passed into a not unhappy oblivion, the fundamental idea of an open universe in which uncertainty, choice, hypotheses, novelties and possibilities are naturalized will remain associated with the name of James; the more he is studied in his historic setting the more original and daring will the idea appear. . . . Such an idea is removed as far as pole from pole from the temper of an age whose occupation is acquisition, whose concern is with security, and whose creed is that the established economic regime is peculiarly "natural" and hence immutable in principle.

JOHN DEWEY

No one yet has succeeded, it seems to me, in jumping into the centre of your vision. . . . That it is the philosophy of the future, I'll bet my life.

WILLIAM JAMES TO JOHN DEWEY

As the philosophy of Spencer had reigned supreme in the heroic age of enterprise, so pragmatism, which, in the two decades after 1900, rapidly became the dominant American philosophy, breathed the spirit of the Progressive era. Spencer's outlook had been the congenial expression of a period that looked to automatic progress and laissez faire for its salvation; pragmatism was absorbed into the national culture when men were thinking of manipulation and control. Spencerianism had been the philosophy of inevitability; pragmatism became the philosophy of possibility.

The focus of the logical and historical opposition between pragmatism and Spencerian evolutionism was in their approach to the relationship between organism and environment. Spencer had been content to assume the environment as a fixed norm — a suitable enough position for one who had no basic grievance against the existing order. Pragmatism, entertaining a more positive view of the activities of the organism, looked upon the environment as something that could be manipulated. It was by way of the pragmatists' theory of mind in relation to environment that the old outlook was controverted.

Pragmatism resulted from criticism not only of Spencerian evolutionism but of many other intellectual tendencies. It was, certainly in its inception, by no means an essentially social philosophy. At the risk, then, of considerable oversimplification, and with this caution, it will be of use to the main purpose of this study to look briefly into the relation of the two leading pragmatists to Spencerianism and to the general social outlook that was being so roughly challenged in the days of pragmatism's ascendancy.

Spencerian philosophical doctrines were just making their way in the United States when other currents began to stir. Before the *Synthetic Philosophy* was completed, a Hegelian movement was under way and pragmatism was well advanced in its formative stages. In 1867 William T. Harris, unable to persuade the editors of the *North American Review* to publish a critical piece on Spencer, launched his own *Journal of Speculative Philosophy*. The Hegelian idealism of Harris and the St. Louis School gained with surprising rapidity and soon acquired the stature of an active competitor of Spencerianism and the old Scottish philosophy. In spite of the gulf that separated them in philosophic doctrine, Hegelianism and Spencerianism, as they were usually interpreted in America, shared a common social conservatism.[1] The same cannot be said categorically of pragmatism, which displayed much greater flexibility in its potentialities for social thought.

Although profoundly influenced by Darwinism, the pragmatists soon departed sharply from prevailing evolutionary thought. Hitherto evolutionism, because of its identification with Spencer, had been blown up into a cosmology. The pragmatists turned philosophy from the construction of finished metaphysical systems to an experimental study of the uses of knowledge. Pragmatism was an application of evolutionary biology to human ideas, in the sense that it emphasized the study of ideas as instruments of the organism. Working primarily with the basic Darwinian concepts — organism, environment, adaptation — and speaking the language of naturalism, the pragmatic tradition had a very different intellectual and practical issue from Spencerianism. Spencer had apotheosized evolution as an impersonal process, the omnipotence of circumstances and the environment, the helplessness of man to hasten or deflect the course of events, the predetermined development of society in accordance with a cosmic process toward a remote but comfortable Elysium. By defining life and mind as the correspondence of inner to outer relations, he had portrayed them as essentially passive agencies. The social counterpart of this approach was his gradualistic fatalism.[2] The pragmatists, beginning without any special interest in ulterior social consequences, at first approached the uses of ideas in an individualistic vein, but in time drifted toward a socialized philosophical theory in the form of Dewey's instrumentalism. The development and spread of pragmatism broke Spencer's monopoly on evolution, and showed that the intellectual uses of Darwinism were more complex than Spencer's followers had thought. The pragmatists' most vital contribution to the general background of social thought was to encourage a belief in the effectiveness of ideas and the possibility of novelties — a position necessary to any philosophically consistent theory of social reform. As Spencer had stood for determinism and the control of man by the environment, the pragmatists stood for freedom and control of the environment by man.

To find the beginnings of pragmatism and its critique of the older evolutionism, one must look beyond James and Dewey to Chauncey Wright and Charles Peirce. Although Wright and Peirce were essentially technical philosophers, they were also critics of established social thinking, including the Spencerian apparatus. It was their experimental criticism that James broadened into a humanistic philosophy, and it was a related philosophical outlook that in the hands of Dewey became both a social theory and a social influence.

Chauncey Wright (1830–75) was the intellectual leader of the Metaphysical Club, founded by Peirce in the 1860's and attended by James, Fiske, the younger Holmes, and a few other Cambridge intellectuals. The best of Wright's philosophical work evidently emerged in conversation at such gatherings, but he also had access to the public as a critic for the *North American Review* and the *Nation*. Both James and Peirce were stimulated by his hard-headed, empirical way of thinking.[3]

Wright was probably the first American thinker to publish a thoroughgoing critique of Spencer from a naturalistic point of view. Steeped in the writings of John Stuart Mill, Wright objected to the popular tendency to classify Spencer as a positivist. He condemned him as a second-rate metaphysician presuming to deal in ultimate truths, and charged him with a misleading devotion to useless abstractions. In this essay he declared:

> Nothing justifies the development of abstract principles but their utility in enlarging our concrete knowledge of nature. The ideas on which mathematical Mechanics and the Calculus are founded, the morphological ideas of Natural History, and the theories of Chemistry are such working ideas, — finders, not merely summaries of truth.[4]

As an alternative to the finality of Spencer's view of scientific knowledge, Wright believed in the probability of novelties in the universe, a conviction based upon a strict interpretation of the inductive character of scientific laws.[5] It is possible for " accidents " or novelties to arise which are not predictable from our knowledge of their antecedents — for example, the evolution of self-consciousness, or the application of the voice to social communication.

Charles Peirce (1839–1914) was more inclined to system-building than either James or Wright, but his orientation was also scientific. The son of Benjamin Peirce, the eminent Harvard mathematician, Charles Peirce attained distinction in his own right as a mathematician, astronomer, and geodesist. Primarily interested in logical theory, and particularly in the problem of induction, Peirce viewed scientific laws as statements of probabilities rather than invariable relationships; in particular instances the facts are certain " to show irregular departures from the law." [6] A thoroughly consistent evolutionist, Peirce argued, must regard the laws of nature themselves as the results of evolution, and hence as limited rather than absolute. There exists, he concluded, " an element of indeterminacy, spontaneity, or absolute chance in nature." [7] Consequently he had no sympathy for Spencer's attempt to deduce evolution from a mechanical principle (persistence of force) instead of explaining the emergence of mechanical principles from evolution. Also he pointed out that, since the law of conservation of energy is equivalent to the proposition that all operations governed by mechanical laws can be reversed, continued growth cannot be explained by these laws.[8] Peirce's stringent criticisms of Spencer first turned James from the *Synthetic Philosophy* to a more experimental approach, and Peirce's epoch-making essay, " How to Make Our Ideas Clear," published in the *Popular Science Monthly* in 1878, first formulated the practical criterion of the meaning of ideas which James later expanded into the pragmatic theory of truth.[9]

II

William James was the first great beneficiary of the scientific education emerging in the United States during the 1860's and 1870's. Trained at Harvard's Lawrence Scientific School by Eliot, Wyman. and Agassiz, James was distinguished from most of his contemporaries by an understanding of scientific method coupled with a broad streak of mysticism, an acute moral and aesthetic sensitivity. This may be traced in part to the influence of his father, the elder Henry James who was a Swedenborgian.[10] Personal factors, emotion, and temperament, loom unusually large in the formation of James's thought. In 1869–70 he underwent a severe emotional depression in which he almost lost his will to live; from it he emerged with a highly intellectualized solution, a passionate belief in freedom of the will.[11] The revolt against " the monistic superstition under which I had grown up " [12] — a dominant motif in his thought — led him into

rebellion against all " block-universe " philosophies, all systems which were finished and executed, impervious to chance
or choice.   Primarily, his revolt against the prevailing philosophies, Spencerianism and Hegelianism, was a revolt
against the moral and aesthetic bleakness of hidebound philosophical systems.[13]   This was the source of the pluralism he
preached.   " It surely is a merit in a philosophy to make the
very life we lead real and earnest," he wrote.   " Pluralism, in
exorcising the absolute, exorcises the great de-realizer of the
only life we are at home in, and thus redeems the nature of
reality from essential foreignness." [14]

James's thought is usually looked upon as a reaction from
the absolute idealism expounded by his Harvard colleague,
Josiah Royce, and in the writings of his later years this reaction is particularly clear; but the basic trend of his thinking, which was well established before he became acquainted
with Royce, was visible in his work at the same time that the
St. Louis Hegelians were beginning to spread their creed.
James's original impulse was in a large measure also a reaction from Herbert Spencer.   The years when Spencer's philosophy was winning attention were the formative years of
James's intellectual life.   He read *First Principles* in the early
1860's and soon enrolled in the ranks of Spencerian converts.   So infatuated was he with the intellectual revolution which Spencer had apparently accomplished that he felt
" spiritually wounded " when Charles Peirce attacked the
master in his presence.[15]   Yet Peirce's arguments prevailed;
James soon began to criticize Spencer himself, and by the
middle 1870's he was roundly contemptuous of Spencer's
ponderous system.   Although he was using the *Principles of
Psychology* in his teaching at Harvard, he urged his students
to criticize Spencer's reasoning, and for almost three decades
he offered courses in which the Englishman served as a whip-
ping-boy.[16]   While he considered *First Principles* a museum
of logical confusions, the emotional core of his reaction to
Spencer is suggested by James's complaint that Spencer's mind
was " so fatally lacking in geniality, humor, picturesqueness,
and poetry; and so explicit, so mechanical, so flat in the pano-

rama which it gives to life." [17] James was overcome by the "awfully monotonous quality " in Spencer, remarking that "one finds no twilight region in his mind, and no capacity for dreaminess and passivity. All parts of it are filled with the same noonday glare, like a dry desert where every grain of sand shows singly, and there are no mysteries or shadows." [18] In his *Pragmatism* he paused to object to Spencer's "dry school-master temperament . . . his preference for cheap makeshifts in argument, his lack of education even in mechanical principles, and in general the vagueness of all his fundamental ideas, his whole system wooden, as if knocked together out of cracked hemlock boards." [19] The margins of James's copy of *First Principles* he decorated with such remarks as "absurd," "trebly asinine," "curse his metaphysics," and "damned scholastic quibble." He made fun of the vacuity of Spencer's fundamental principles, declaring Spencer's use of the persistence of force to be "vagueness incarnate," and suggesting ridiculous examples for his principle of the rhythm of motion, such as people going upstairs and downstairs, intermittent fevers, and cradles and rocking chairs. Spencer's definition of evolution he parodied in his lectures in these terms: "Evolution is a change from a no-howish untalkaboutable all-alikeness to a somehowish and in general talkaboutable not-all-alikeness by continuous stick-togetherations and somethingelseifications." [20]

It seems clear that James's objection to Spencer arose partly because James was in search of a philosophy that would acknowledge active human effort in the bettering of life. In *Pragmatism* James objected to Spencer's finished deterministic philosophy because of the "disconsolateness of its ulterior pratical results." [21] "When the whole training of life is to make us fighters for the higher," he wrote on the last page of his well-annotated copy of *First Principles,* "why should it be extraordinary or wrong to protest against a philosophy the acceptance of which is acceptance of the defeat of the higher? " [22] Symptomatic of James's approach to philosophy was his recurrent interest in the problem of evil, so evident in his answers to Royce   His desire to reject all philosophies that denied the existence of evil or minimized its practical significance can be seen in his attacks upon absolutism. He cited with approval an indictment of philosophers drawn up by Morrison I. Swift, an anarchist writer, for their cool neglect of the ills of society.[28] His essay on " The Dilemma of Determinism " hinged upon the necessity of validating moral judgments. The idea of chance, which interested Peirce as a logician, was significant to James as a moralist. Determinism, James declared, insists that possibilities which fail to get realized are no possibilities at all, but illusions; it affirms that nothing in the future can be ambiguous, not even human volitions. But if there is no such thing as chance, moral judgments are pragmatically meaningless.

> Calling a thing bad means, if it means anything at all, that the thing ought not to be, that something else ought to be in its stead. Determinism, in denying that anything else can be in its stead, virtually defines the universe as a place in which what ought to be is impossible, — in other words, as an organism whose constitution is afflicted with an incurable taint, an irremediable flaw.[24]

Determinism is consistent only with the direst pessimism or a romantic mood of resignation. But if moral judgments are to be effective there must be some minimum of uncertainty in the universe; this does not necessitate a completely haphazard world, but only one in which there are occasional choices. The necessity of retaining choices remains, even though one's dream of universal fatalism be as optimistic as that of Spencer, who believes in the ultimate advent of a peaceful millennium. Even if it is true, as Spencer says, that no preference can succeed unless it is in harmony with the ultimate triumph of peace, justice, and sympathy, " we are still free to decide *when* to settle down on the equitable and peaceful basis." Until it is finally revealed with certainty what shall succeed, we are all free to try for our own preferences.[25]

In 1878 the *Journal of Speculative Philosophy* published an article by James entitled " Remarks on Spencer's Definition of Mind as Correspondence," in which the lines of his later thought are clearly foreshadowed. The article shows also how much more dynamic than Spencer's was James's understanding of the implications of Darwinism for psychology. Spencer, in defining the mind in terms of adjustment, leaves out the greater part of what is usually considered the mental life. Spencer defines life as the adjustment of inner to outer relations and looks upon mind and cognition as aspects of that adjustment. He forgets, according to James, all noncognitive elements in mind, all sentiment and emotion. He plays down or ignores entirely the element of *interest* in the organism which is essential to the whole process of cognition. He defines intelligent mental reactions as those that minister to survival by arranging internal relations to suit the environment, but the critical factor in the cognitive situation, the *desire* for survival or welfare, is a subjective element which he leaves out. The idea of correspondence between inner and outer relations, to be made meaningful as the criterion of mental acts, must be qualified by some subjective or teleologic reference. Furthermore, the idea that mind ministers to survival alone cannot explain the full range of higher cultural activities which have no survival value. The knower, James concluded,

. . . is not simply a mirror floating with no foothold anywhere, and passively reflecting an order that he comes upon and finds simply existing. The knower is an actor, and coefficient of the truth on one side, whilst on the other he registers the truth which he helps to create. Mental interests, hypotheses, postulates, so far as they are bases for human action — action which to a great extent transforms the world — help to *make* the truth which they declare. In other words, there belongs to mind, from its birth upward, a spontaneity, a vote. It is in the game, and not a mere looker-on; and its judgments of the *should-be*, its ideals, cannot be peeled off from the body of the cogitandum as if they were excrescences, or meant, at most, survival.[26]

In his *Principles of Psychology*, which appeared in 1890, James continued this line of thought. There he made a sharp break with the traditional view of mind as a quiet cognitive organ, and criticized post-Darwinian psychology for its neglect of the active role of the mind.[27] It had become habitual, he complained, to speak as if the mere body that owns the

brain has interests, to treat the body's survival as an absolute
end without reference to any commanding intelligence. In
this bare physical view, the reactions of an organism cannot
be considered useful or harmful; it can only be said of them
that if they occur in certain ways survival will incidentally
be their consequence:

> But the moment you bring a consciousness into the midst, survival
> ceases to be a mere hypothesis. No longer is it " *if* survival is to occur,
> then so and so must brain and other organs work." It has now become
> an imperative decree: " Survival *shall* occur, and therefore organs *must*
> so work! " *Real* ends appear for the first time upon the world's
> stage. . . . Every actually existing consciousness seems to itself at any
> rate to be a *fighter for ends,* of which many, but for its presence, would
> not be ends at all. Its powers of cognition are mainly subservient to
> these ends, discerning which facts further them and which do not.[28]

The doctrine — or method — of pragmatism, taken with acknowledgment from Peirce, was a projection of this approach
to the test of knowledge. A world in which theories are
experimental instruments rather than answers, and in which
truth " happens to an idea " [29] and can be made by the
knower, was alone coherent with the unfinished universe
James chose to believe in.

In 1880 James made one of his rare ventures into social
theory when he published in the *Atlantic Monthly* an article
on " Great Men, Great Thoughts, and the Environment." [30]
Using Spencer and his disciples as a foil, James raised the
question: What causes communities to change from generation to generation? With Walter Bagehot, whose *Physics
and Politics* he greatly admired, James believed the changes
were the result of innovations by unusual or outstanding
individuals, playing the same role in social change as variations in Darwin's theory of evolution; such persons are selected by society and elevated into positions of influence because of their adaptability to the social situation into which
they happen to be born. The Spencerians had attributed
social changes to geography, environment, external circumstances — in brief, to everything except human control.
They had assumed the existence of a universal web of causa-

tion, in which the finite human intellect becomes hopelessly entangled. Spencer, in his *Study of Sociology*, attributed everything to prior conditions, a process which, when pushed further and further back, becomes circular and yields nothing of value in social analysis. It offers nothing but the omnipotence of circumstances, in much the same way as the Oriental answers every question with " God is great." This evolutionary philosophy has no explanation for the patent fact that great men change the course of social development. The significant details of their careers cannot be predicted by or attributed to the vague complex of factors which in the Spencerian philosophy is invoked by the term " environment." There was no convergence of sociological pressures which demanded that William Shakespeare be born at Stratford-on-Avon in 1564. The most that sociology can predict is that *if* a great man of such nature appears under certain circumstances he will affect society in such and such a way; but that he does affect it should not be denied. The great man is himself part of the environment of everybody else.

Spencer's impersonal view of history is a brand of oriental fatalism, " a metaphysical creed and nothing else. It is a mood of contemplation, an emotional attitude rather than a system of thought "; and in its neglect of spontaneous variations in human thinking and their effect upon society, it is " an absolute anachronism reverting to a pre-Darwinian type of thought." [31] In this essay James seems to be out-individualizing the individualists, but in the larger context of his thought it appears that his main concern was to redeem spontaneity and indeterminacy from the oppressive causal network of Spencerian social evolution. Without spontaneity, without some possibility that the individual may in a measure alter the course of history, there is no chance for betterment of any kind, and the whole romance of struggle with its attendant alternatives of triumph or failure is banished. As James declared in a subsequent article, " there is a zone of insecurity in human affairs in which all the dramatic interest lies. The rest belongs to the dead machinery of the stage." That life should be deprived of its dramatic interest by a scheme of universal causality was an intolerable thought, " the most pernicious and immoral of fatalisms." [82]

Unlike Dewey, his successor in the pragmatic tradition, James was guilty of only the remotest interest in systematic or collective social reform. One expression of his fundamental individualism [83] is the fact that while he was from time to time interested in current events — as an anti-imperialist, a Dreyfusard, a Mugwump — he had no sustained interest in social theory as such. He always dealt with philosophical problems in individual terms. When he wished to illustrate the problem of evil he chose a spectacularly brutal murder rather than wars or slums.[34] Although he was casually interested in moderate reform, he had been brought up in the brand of liberalism displayed by the *Nation,* to which he declared he owed his " whole political education." [85] He considered Godkin a fount of political wisdom [86] and Spencer's political and ethical theory, in spite of its " hardness and inflexibility of tone," infinitely superior to his abstract philosophy.[87] He thought Spencer inconsistent in attempting to be faithful at once to the old English tradition of individual liberty as embodied in *Social Statics* and to the theory of universal evolution, in the workings of which the individual's interests are often harshly overridden.[88] He was never capable of the sternness of spirit that marked more consistent old-school liberals of Godkin's stripe. He could not take alarm at the activities of labor, even in the heated days of 1886, when he wrote to his brother that labor troubles " are a most healthy phase of evolution, a little costly, but normal, and sure to do lots of good to all hands in the end " — excepting, of course, the anarchist riots in Chicago, which were " the work of a lot of pathological Germans and Poles." [89]

In his later years, after the winds of social criticism had been blowing in America for some time, James viewed the rise of collectivism with satisfaction and found a means of reconciling it with his characteristic emphasis on individual activity. " Stroke upon stroke, from pens of genius," he wrote to Henry James in 1908, after reading G. Lowes Dickinson's

Justice and Liberty, " the competitive regime so idolized seventy-five years ago seems to be getting wounded to death. What will follow will be something better, but I never saw so clearly the slow effect of [the] accumulation of the influence of successive individuals in changing prevalent ideals."[40] In 1910 he openly expressed his belief " in the reign of peace and the gradual advent of some sort of socialistic equilibrium." [41] Yet in one of his lectures over a decade before, he had, in a more characteristic Jamesean vein, advised a very modest appraisal of the meaning of collectivism for the inner quality of human life:

> Society has . . . undoubtedly got to pass toward some newer and better equilibrium, and the distribution of wealth has doubtless slowly got to change; such changes have always happened, and will happen to the end of time. But if, after all that I have said, any of you expect that they will make any *genuine vital difference,* on a large scale, to the lives of our descendants, you will have missed the significance of my entire lecture. The solid meaning of life is always the same eternal thing, — the marriage, namely, of some unhabitual ideal, however special, with some fidelity, courage, and endurance; with some man's or woman's pains. — And, whatever or wherever life may be, there will always be the chance for that marriage to take place.[42]

III

" A *real school,* and *real Thought.* Important thought, too! " [43] This was the reaction of William James to the group of philosophers and educators gathered around John Dewey at the University of Chicago in the early 1900's. Dewey's indebtedness to James and James's approval of Dewey indicate the essential continuity of the pragmatic school. When Dewey first read James's *Principles of Psychology* he was still under the sway of the Hegelianism of George Sylvester Morris, with whom he did his graduate work at Johns Hopkins and his earliest teaching at Michigan. But James's psychology altered the whole trend of his thinking, and James's approach to mental life became a vital strain in his philosophy.[44] With James, Dewey preached the effectiveness of intelligence as an instrument in modifying the world; but to the philosophical argument he brought an unusually strong consciousness of its social import and an urgent sense of the social responsibility of the philosopher. The instrumentalist view of the creative character of intelligence he associated with experimentalism in social theory — a sharp contrast to the conservatism everywhere prevalent in 1882 when Dewey came to Baltimore to begin his graduate study.[45]

Dewey's interpretation of the act of thought is more than a simple extension of Darwinism, but it is biological in its orientation.[46] Thinking is not a series of transcendent states or acts interjected into a natural scene. Knowledge is a part of nature, and its end is not mere passive adjustment but the manipulation of the environment to provide " consummatory " satisfactions. An idea is a plan of action rooted in the natural impulses and responses of the organism. The " spectator theory of knowledge " is pre-Darwinian.[47] " The biological point of view commits us to the conviction that mind, whatever else it may be, is at least an organ of service for the control of environment in relation to the ends of the life process." [48] Dewey joined this view of mental activity to a general criticism of the conservative outlook. As he remarked in 1917:

The ultimate refuge of the standpatter in every field, education, religion, politics, industrial and domestic life, has been the notion of an alleged fixed structure of mind. As long as mind is conceived as an antecedent and ready-made thing, institutions and customs may be regarded as its offspring.[49]

Associated with Dewey's belief in the potentialities of intelligence was his insistence that it operates within a series of objectively " indeterminate " situations. It is from the indeterminateness of the situations, from the element of contingency in nature, that the discriminating intellect derives its special significance. The significance of morals and politics, of religion and science, have " their source and meaning in the union in Nature of the settled and the unsettled, the stable and the hazardous." Without this union there can be no such things as " ends," either in the form of human consummations or as purposes. " There is only a block universe, either something ended and admitting of no change, or else a predestined march of events. There is no such thing as fulfillment where there is no risk of failure, and no defeat where there is no promise of possible achievement." [50] Although suggestive of James, this conception is probably closer to the earlier views of Peirce and Wright, for Dewey shied clear of James's assertion of freedom of the will.[51]

In 1920, in his *Reconstruction in Philosophy*, Dewey made a powerful argument for a practical emphasis in philosophy and urged that philosophers shift their attention from the sterile aspects of epistemology and metaphysics to politics, education, and morals. Appended as it was to an acute historical analysis of the division between thought and action, the volume stood as his most noteworthy statement on the theme. The social emphasis, however, was deeply rooted in Dewey's career. A decade before, he had predicted that philosophy would become, among other things, " a moral and political diagnosis and prognosis," and as early as 1897 he had stated his social view of the problem of knowledge.[52]

From his early years, when he became acquainted with Comte's writings, social philosophy had occupied a prominent place among Dewey's concerns.[53] In 1894 he published a review of Ward's *Psychic Factors of Civilization* and Kidd's *Social Evolution* which enables the reader to place Dewey's thinking in relation to the problem of biological sociology. Dewey approved of Ward's effort to overthrow mechanical Darwinism in sociology by means of a theory of psychic activity. Spurred by James's psychology, however, Dewey was critical of Ward's old-fashioned allegiances in this field, and he pointed out that Ward's psychology was an inadequate instrument for his social theory. In his version of psychic activity, the critical point in his whole sociology, Ward worked with an outmoded pleasure-pain psychology, not far advanced from the sensationalism of Locke. From such passive states of feeling as pleasure and pain sensations, Ward attempted to derive action. His psychology would be better founded if it rested upon the primary fact of *impulse*, the positive motivation of the organism, " just the fact needed to give firm support to his main contentions." In his criticism of

Ward's psychology, Dewey was rearguing in a general way the case that James had made against Spencer sixteen years before. With Ward's criticism of laissez faire Dewey agreed, and he also agreed that " the biological theory of society needs reconstruction from the standpoint of the significance of intellect, emotion, and impulse." His differences with Ward were over means, not ends.

Dewey's criticisms of Kidd were more fundamental. While he admitted that the elimination of conflict from society is " a hopeless and self-contradictory ideal," he still believed it possible to direct the struggle to eliminate waste. He argued that Kidd's belief in the continual sacrifice of the individual to the conditions of progress showed a hopeless confusion about the relation of ends and means. In Kidd's scheme the individual forever sacrifices himself to the welfare of future generations, but since the individuals of future generations do the same, the process never reaches any consummation in human satisfactions. Man always sacrifices toward an end which is by definition never attained.[54] It was the *reductio ad absurdum* of the philosophy of progress.

Dewey's skepticism about laissez faire was a logical consequence of his experimentalism. Far from agreeing with Spencer's contention that meddling in social affairs is a barrier to knowing them, he insisted that direct participation in events is necessary to genuine understanding.[55] No universal proposition can be laid down to determine the functions of a state. Their scope is something for experimental determination.[56] The general reaction from laissez faire in practical policy commanded his sympathy, but Dewey deplored the absence of a coherent alternative theory and the general penchant for working on a vague belief that something must be done.[57] His emphasis upon the function of education in social change, reminiscent of the earlier proposals of Lester Ward, derives in part from a sense of need for guidance.[58]

Dewey's ethical speculation was an attempt to bring order out of the moral confusion caused by the apparently conflicting aims of morality and science, and it is especially significant that he attributed the development of instrumentalism largely to the stimulation afforded by this problem.[59] In an early article in *The Monist* (1898) which showed marks of his Hegelian background, Dewey rejected Huxley's distinction between the cosmic process and the ethical process. While disputing Huxley's dualistic approach to the problem, Dewey did not doubt the validity of Huxley's analogy between the ethical and the horticultural process. " The ethical process, like the activity of the gardener, is one of constant struggle. We can never allow things simply to go on of themselves. If we do, the result is retrogression." But what is the significance of this apparent opposition between the ethical process and the cosmic process, in the light of our idea of the evolutionary process as a whole? At this point Dewey argued that Huxley failed to realize that the conflict is not one in which man is pitted against his entire natural environment, but one in which man modifies one part of the environment with relation to another. He does not work with anything that is totally alien to his entire environment. The gardener may introduce foreign fruits or vegetables into a particular locale, and he may assist their growth by conditions of sunlight and moisture unusual on his particular plot of ground; " but these conditions fall within the wont and use of nature as a whole."

Huxley recognized that the survival of the fittest in respect to the existing conditions is different from the survival of the ethically best. Yet must not the conditions be interpreted as a whole complex, including " the existing social structure with all the habits, demands, and ideals which are found in it? " Under such an interpretation the fittest would in truth be the best. The unfit would be practically equivalent to the antisocial, but not to the physically weak or the economically dependent. The dependent classes in society may be quite " fit " when measured by the whole of the environment. The prolongation of the period of dependency in man (Fiske's infancy theory) has developed foresight and planning and the bonds of social unity; the care of the sick has taught us how to protect the healthy. What was fit among the carnivora is not fit among men. Man lives in such a changing and progressive environment that in his case it is flexibility, readiness to adjust to the conditions of the morrow as well as the present, that constitutes fitness. As the meaning of environment changes, the meaning of the struggle for existence changes also. The biological promptings of self-assertion have potentialities for good as well as evil. The essence of the human problem is controlled foresight — ability to maintain the institutions of the past while remaking them to suit new conditions; in short, to maintain a balance between habits and aims. The term "selection" can mean not only that one form of life, one organism, is selected at the expense of another, but also that various modes of action and reaction are selected by an organism or a society because of their superiority over other modes. Society has its own mechanisms, public opinion and education, to select the modes it finds most suitable. There is, then, no bifurcation between the ethical process and the cosmic process. The difficulty has been created by static interpretations of biological functions and their application out of context to the unique and dynamic conditions of human environment. It is not necessary to go outside of Nature to find warrant for the ethical process; one need only recognize the natural situation in its totality.[60]

In 1908 Dewey and his former colleague James H. Tufts brought out a textbook in ethics whose contents were radically different from the abstract homilies characteristic of the past. The treatment of ethical principles was made ancillary to social issues of the day; such problems as individualism and socialism, business and its regulation, labor relations and the family, were given prominent place. Dewey's contribution included a short section sharply criticizing the crude assimilation of Darwinism to ethical theory, and took a position on the "natural" aspect of competition not unlike Kropotkin's.[61]

Indeed, it is as part of a maturing social criticism that the historical position of pragmatism can best be understood. This is in accord with Dewey's own conception of the place of the pragmatic tradition in American culture. Dewey repeatedly denied that pragmatism, with its interest in what James called the " cash-value " of ideas, was an intellectual equivalent of American commercialism or an abject apology for the acquisitive spirit of a business culture. He reminded its critics that it was James who protested against the excessive American worship of " the bitch-goddess SUCCESS." [62] Hostile to all absolutist social rationalizations, conservative or radical, instrumentalism has varied in social content from the Progressive era to the days of the New Deal; but what is most important in its history is its association with social consciousness and its susceptibility to change.

The social orientation which was always present in Dewey's thought, in bold relief against the individualism of William James, illustrates the changing potentialities of a somewhat similar philosophical standpoint in different eras. It was partly a difference in personal history. James came from a family with a comfortable inherited fortune that provided him with a Harvard education, travel, social status, and the advantage of a long period of maturation free of financial problems. Dewey, the son of a small business proprietor in Burlington, Vermont, was thrown largely upon his own resources. The social emphasis acquired by instrumentalism has, however, a larger significance. Dewey, who was born in the year when *The Origin of Species* appeared, survived James for a period in which two generations came to maturity, and social criticism among men of academic standing became respectable. The beginnings of the Progressive era, moreover, coincided with the growth and spread of Dewey's ideas — the same period in which James himself thought he saw the competitive regime " getting wounded to death "; and it is easy to see Dewey's faith in knowledge, experimentation, activity, and control as the counterpart in abstract philosophy of the Progressive faith in democracy and political action. It is not far from Croly's appeal to his countrymen to think in terms of purpose rather than inevitable destiny, or from Lippmann's assertion that " we can no longer treat life as something that has trickled down to us," to Dewey's appeal for an experimental approach to social theory. If Dewey's belief in the efficacy of intelligence and education in social change was justified, his own philosophy was more than a passive reflection of the transformation in American thought. The sight of a distinguished philosopher occupied with the activities of third parties, reform organizations, and labor unions provided a measure of some of the changes that have taken place on the American intellectual stage since the days when Fiske and Youmans were dramatizing Spencer for an enchanted audience.

Chapter Eight

Trends in Social Theory
1890–1915

Except for a possible reversion to a cultural situation strongly characterized by ideas of emulation and status, the ancient racial bias embodied in the Christian principle of brotherhood should logically continue to gain ground at the expense of the pecuniary morals of competitive business.

THORSTEIN VEBLEN

The Struggle for Existence is another of these glimpses of life which just now seems to many the dominating fact of the universe, chiefly because attention has been fixed upon it by copious and interesting exposition. As it has had many predecessors in this place of importance, so doubtless it will have many successors.

CHARLES HORTON COOLEY

Evolution had a profound impact upon psychology, ethnology, sociology, and ethics but it failed to work a similar transformation in economics. William Graham Sumner was alone among those who could be considered economists in his attempt to assimilate evolution to the traditional concepts of political economy. During the 1870's and 1880's, when he was working out the fundamentals of his social philosophy, most other economists were far more traditional in their thinking.

The most plausible explanation of the inflexibility of the received political economy is that its spokesmen were satisfied that there was little their science could learn from biology. The accepted function of political economy, as taught in American colleges and propagated in the forums of opinion, was apologetic. It had always been an idealized interpretation of economic processes under the competitive regime of property and individual enterprise; violations of the set pattern had been discouraged as infringements of natural law As Francis Amasa Walker said of the laissez-faire principle in the period before the founding of the American Economic Association, " Here [in the United States] it was not made the test of economic orthodoxy, merely. It was used to decide whether a man were an economist at all." [1]

The common failure of orthodox economists to embrace social Darwinism as preached by Sumner — so obviously adapted to the function of their science as they conceived it — was only incidentally due to the unsettled status of evolution in its relation to religious beliefs. More important was the fact that classical economics already had its own doctrine of social selection. Since it had been one of the great figures of the classical economic tradition who had led Spencer, Darwin, and Wallace toward their evolutionary theories, the economists might have had some justification for proclaiming that biology had merely universalized a truth that had been in their possession for a long time.

A parallel can be drawn between the patterns of natural selection and classical economics,[2] suggesting that Darwinism involved an addition to the vocabulary rather than to the substance of conventional economic theory. Both assumed the fundamentally self-interested animal pursuing, in the classical pattern, pleasure or, in the Darwinian pattern, survival. Both assumed the normality of competition in the exercise of the hedonistic, or survival, impulse; and in both it was the " fittest," usually in a eulogistic sense, who survived or prospered — either the organism most satisfactorily adapted to its environment, or the most efficient and economic producer, the most frugal and temperate worker. Here, it should be added, economics was the better suited to a kindly interpretation of the status quo since it accepted the present environment as a natural datum, while conscientious and perceptive followers of Darwin saw that the " fittest " might be understood to be fit to inferior and degrading surroundings. Veblen, writing in 1900, found that " identification of the categories of normality and right gives the dominant note of Mr. Spencer's ethical and social philosophy, and that later economists of the classical line are prone to be Spencerians." [3] Further, both classical economics and natural selection were doctrines of natural law. At this point, once again, classical economics was more conducive to intellectual stability, because its concept of equilibrium was Newtonian and hence static,[4] whereas a dynamic theory of society raised possibilities of an unsettled world.

The conception of the pressure of population upon subsistence, so important in the historic connection between biology and political economy, not only played its part in the doctrine of Malthus but was closely related to the classic wage-fund doctrine. According to the wage-fund theory, which was popular among extreme proponents of laissez faire in the United States,[5] labor is paid out of a capital fund which is fixed at any given time; and the average wage of the laborers is determined by the ratio of the number of workers seeking employment to the amount of the wage-fund. According to the logic of the wage-fund theory, neither legislative regulation nor any action of the laborers could alter this state of affairs, and a policy of strict acquiescence was indicated. Competition was generally considered the perfect means of distributing wealth. According to this doctrine the increase in numbers among the working class pressed upon the limited wage-fund with the same inexorability as the total population on the means of subsistence. This doctrine, said Walker, " was not a little favored by the fact that it afforded a complete justification for the existing order of things respecting wages." [6] After the publication in 1876 of his study of *The Wages Question*, however, its prestige declined rapidly.

The content of American economic thought was not quickly changed. In the post-bellum decades the most popular college textbook was a revised version of the Rev. Francis Wayland's *The Elements of Political Economy* — originally written in 1837. It was from this book that both Sumner and Veblen learned their college economics. Wayland's original aim had been to present a methodical restatement of the doctrines of Smith, Say, and Ricardo. Wayland and such other representatives of the classic tradition as Francis Bowen, Arthur Latham Perry, and J. Laurence Laughlin were in substantial agreement on the premises of economic science. Man is a creature of desires, universally motivated by self-interest; the mechanism of competition, if free and fair, transmutes the self-seeking of the economic man into deeds that work for " the greatest good of the greatest number." But this machinery is delicate and must be permitted to operate under " normal " conditions and must not be overridden by government interference; to enjoy the fruits of an inherently beneficent natural economic law, men must permit it to operate unhampered; they must be industrious, frugal, temperate, and self-reliant; self-help, and not a weak recourse to state intervention, is the way of economic salvation.[7] It had required no hard labor on Sumner's part to fit this pattern to Darwinian individualism.

Developments in the middle 1880's indicated that traditional economic thought was losing its grip on younger scholars, partly through the influence of the German historical school. Richard T. Ely, fresh from his graduate education at Halle and Heidelberg, published an essay on "The Past and the Present of Political Economy," in which he attacked classical economics for its dogmatism and simplicity, its blind faith in laissez faire, and its belief in the adequacy of self-interest as an explanation of human conduct. Ely praised the historical school as an antidote, arguing that the historical method could not lead to such doctrinaire extremes.

. . . this younger political economy no longer permits the science to be used as a tool in the hands of the greedy and the avaricious for keeping down and oppressing the laboring classes. It does not acknowledge laissez-faire as an excuse for doing nothing while people starve, nor allow the all-sufficiency of competition as a plea for grinding the poor.[8]

In the following year Simon Patten, a midwestern farm boy who had taken his doctorate at Halle, published a critique of *The Premises of Political Economy*, in which he questioned the social utility of unrestricted competition and expressed his dissatisfaction with Malthus, Ricardo, and the wage-fund doctrine. Darwin was generally thought to confirm Malthus' law, Patten said, but in one critical respect Darwin's theory was the exact opposite of Malthusianism. Malthus assumed that man has a definite and unalterable set of attributes; but Darwinism holds that man is pliable and circumstances determine his characteristics. On true Darwinian premises one can assume no such thing as a permanent natural rate of increase; for the human rate of increase would be susceptible to change in accordance with man's surroundings and circumstances.[9]

In 1885 a group of younger economists under Ely's leadership formed the American Economic Association. Its statement of principles read in part as follows:

> We regard the state as an agency whose positive assistance is one of the indispensable conditions of human progress.
>
> We believe that political economy as a science is still in an early stage of its development. While we appreciate the work of former economists, we look not so much to speculation as to the historical and statistical study of actual conditions of economic life for the satisfactory accomplishment of that development.[10]

The membership of the Association as a whole was by no means as critical of tradition as Ely, who was himself not a drastic opponent of the principle of competition.[11] Rather, this statement was expressive of a growing discontent with the simple dogmatism of conventional apologetics. Although Darwinian science meant somewhat more to the young rebels than to most spokesmen of orthodoxy, it was chiefly significant to them as a broad doctrine of change or development; the German historical school rather than Darwinism was their model. "The most fundamental things in our minds," Ely wrote, "were on the one hand the idea of evolution, and on the other hand, the idea of relativity." These things were more important to them than any debate about economic method. "A new world was coming into existence, and if this world was to be a better world we knew that we must have a new economics to go along with it."[12]

II

However limited the impact of Darwinism on economic theory, one could doubtless compile a formidable list of *obiter dicta* in which competition was justified in Sumnerian fashion as a special case of the struggle for existence. Probably the most memorable of such statements was Walker's criticism of Bellamy's case against competition. Walker bridled at the Nationalist argument that the survival of the fittest was a precept of sheer brutality and nothing else. " I must deem any man very shallow in his observation of the facts of life," he commented, " who fails to discern in competition the force to which it is mainly due that mankind have risen from stage to stage in intellectual, moral, and physical power." [18]

Such pronouncements were usually incidental to larger discussions in which Darwinism played no special part. However, two economists, Simon Patten and Thomas Nixon Carver, attempted to go beyond such casual uses of Darwinism and to integrate economics and biology. Patten began his effort with an analysis of the shortcomings of classical economics. Its chief failure was a static conception of human economy. " The environment has so strong an influence over men that their subjective qualities can be neglected. Nature is so niggard and its surplus is so small that no radical change in social relations is possible." Nature seems niggardly until it is shown that the economic environment changes with the changes in men. New classes of men look upon the world in different ways, and the environment they find depends upon their mental characteristics. The laws of a given society are not simply the laws of nature; they are " laws derived from the particular combination of natural forces of which the society makes use."

Modifications of the environment react on men by changing their habits of consumption. Every reduction of cost creates another order of consumption, a new standard of life, which by inducing a new race psychology tends to stimulate new motives in production, new devices, new reductions in cost. This is how a dynamic economy works: progress occurs as a steady upward spiral.[14]

In an essay entitled *The Theory of Social Forces* (1896), Patten enlarged his criticism of prevailing social theory. Current speculation was still dominated by eighteenth-century philosophy and but little affected by the theory of evolution. Here the idea of a changing rather than a static environment held a central position in Patten's system. The theory of goods in economics is in fact study of the environments of organisms. The environment of each organism, being the sum of its economic conditions, changes as these conditions change. In reality there is an indefinite number, a series, of environments. Any given environment, once occupied, is soon filled with struggling beings. "A progressive evolution depends upon the power of moving from one environment to another and thus avoiding the stress of competition." A series of differing environments presents increasingly complex conditions, requiring a new mental evolution for each transition. A progressive nation passes through a complete series of different environments, even though its geographical location does not change. In a progressive evolution the higher type of animal adjusts to a new environment; among lower animals there is a static competition for the existing limited resources. Thus the essence of progress is escape from competition.

Like Ward, Patten distinguished sharply between the biologic and social stages of progress. To this he added his own *ad hoc* distinction between organisms characterized by superior sensory equipment and those having superior motor equipment. Sensory organisms gain clearer ideas of the environment; motor organisms "act with vigor and promptness." In the biologic stage of progress, "beings are pushed into a local environment" in which little thought is required to supply the necessaries of life. The development of motor powers determines who shall survive, and those with inferior motor powers are driven out. Some of these, however, are better fitted to occupy a more general environment in which highly developed sensory powers are of more use. The conquered find a new place to live and create a new society with new requisites for survival. In time the residents of this new society who have the better motor powers will again survive, and those with imperfect motor organization but improved sensory powers are driven once again into a more general environment where new social instincts are needed and a new order is formed. The characteristics of social progress, as distinguished from biological, depend upon this ability to break through from one environment to another.

Applying his concepts to modern society, Patten argued that man has achieved such control over his environment, such development of his sensory faculties, that he has passed out of a pain economy — the primitive economy portrayed in Ricardian economics — into a pleasure economy. The essence of a pleasure economy is not the total absence of pain but the disappearance of fear as a dominant motive. The race slowly loses the instincts of a pain economy and acquires those best suited to the new conditions. In time the pleasure economy's surplus population will be carried off by temptation, disease, and vice, and thus will be bred a race with "instincts to resist extinction by such devices" — a truly superior race of men in a social commonwealth.

In accordance with his general emphasis on the importance of consumption, Patten believed that peoples with varied diets and many wants have a decided advantage over those with simple diet and few wants. "The latter class would require a large area of land to support a given number of persons and would thus be at a disadvantage in an economic contest for survival." Consumption itself becomes a lever in progressive evolution.[15]

Patten failed to make many converts to his novel social theory; and perhaps not without reason, for whatever the merit of his criticisms of classical economics, his own positive theory was more original than substantial. His method was excessively deductive, his distinctions artificial, his expositions exasperatingly vague, his psychology bound by all the limitations of hedonism; but he was an effective teacher and left a lasting impression on many students.[16] In some ways his writings were symptomatic of the Progressive period. He attempted to absorb evolution into economic theory in a thoroughgoing way and to modify classical economics accordingly; and he tried to open new perspectives on the possibilities of a life based upon abundance rather than want.[17]

If Patten sought to find a new place for biology in social and economic theory, it was the task of Thomas Nixon Carver to keep alive the individualism of an earlier day. Appearing chiefly during the Wilsonian period, Carver's ideas sound like a pale echo of the doctrines made familiar by Sumner a quarter of a century before. In a short popular volume entitled *The Religion Worth Having,* Carver preached the life of productive virtue in traditional terms. The best religion, he declared, is that which acts most powerfully as a spur to energy and directs that energy most productively. The religion which best fits men for the struggle to survive will be left in possession of the world, just as the " work-bench " philosophy of life is destined to prevail over the " pig-trough " philosophy. The struggle for existence is primarily a group struggle, but the struggle among individuals continues, and promotes the efficiency of the group in its larger conflict. The group that regulates individualistic competition by rewarding those who strengthen it most and penalizing, through poverty and failure, those who strengthen it least, is the group that will survive. The best method of getting productive work out of men is the selective method of competition, and rewards are best meted out to valuable citizens by means of private property. " The laws of natural selection are merely God's regular methods of expressing his choice and approval," asserted Carver. " The naturally selected are the chosen of God." To help in the essential business of survival the churches should preach obedience to the laws of God through pursuit of the productive life.[18] In his subsequent work Carver continued to defend competition on Darwinian grounds.[19]

More concerned than any other economist with the bearings of post-Darwinian science on economic theory was

Thorstein Veblen. Veblen's conception of the uses of Darwinism in economics was not the most representative of his generation, but in the long run it may be the most enduring. Although Veblen was something of an evolutionary anthropologist, an aspect of his theory best represented in *The Theory of the Leisure Class* (1899) and *The Instinct of Workmanship* (1914), two other features of his achievement may be taken as most relevant to the theme of this study: his assault upon the traditional image of the captain of industry as the " fittest," and his drastic criticism of classical economics in the light of evolutionary science.

In spite of, or perhaps because of, his own association with Sumner at Yale, where he received his doctorate, Veblen had little use for the type of social Darwinism Sumner taught. Veblen once remarked in a review of Enrico Ferri's *Socialisme et Science Positive* that Ferri had shown " in rather more convincing form than is usual with the scientific apologists of socialism " that the equalitarian and collectivist features of socialist theory were not in conflict with the facts of biology. Veblen also expressed cordial approval of Lester Ward's *Pure Sociology*, which he felt had succeeded brilliantly in bringing " the aims and methods of modern science effectively into sociological inquiry." [20]

Veblen's criticisms of the leisure class flatly contradicted Sumner's belief that the well-to-do could be equated with the biologically fittest. In a large sense the greater part of Veblen's work was an inferential criticism of the theoretical scheme of things that equated an individual's productivity with his capacity for acquisition, and the fitness of his character with his pecuniary status. To Sumner accumulation was the reward of personal merit and millionaires were " a product of natural selection "; to Veblen the business class was essentially predatory in outlook and habits. The personal attributes of " the ideal pecuniary man " he described in terms ordinarily reserved for moral delinquents.[21] Where the function of the captain of industry was conventionally considered a productive one, Veblen characterized the methods of a developed business society as an attenuated form of sabotage. Where pecuniary acquisition was regarded as the reward of social service, Veblen distinguished between the productive function of industry as an expression of workmanship, and the partially fraudulent character of business as an expression of salesmanship and chicanery. Where men like Sumner, Walker, and Carver looked upon competition chiefly as a rivalry in productive service, Veblen held that this had been true only in the past when there had been no divorce between business and industry. Competition had once centered about rivalry between producers for industrial efficiency; but when business became supreme over industry it had become chiefly a contest between seller and consumers with a large admixture of fraudulent exploitation.[22]

Not long before the appearance of *The Theory of the Leisure Class*, Veblen wrote a review of Mallock's *Aristocracy and Evolution* which foreshadowed the later development of Veblen's thought. He said his first temptation was to dismiss Mallock's economic argument with the observation that " Mr. Mallock has written another of his foolish books," but he discovered that he could use Mallock's assertions about the value of the captain of industry to enlarge on his own thesis — the unproductive character of the business man.[23]

In *The Theory of the Leisure Class*, Veblen interpreted institutions, individuals, and habits of thought as results of selective adaptation; but his views on the types of character selected for dominance in business society were hardly congenial with the Spencer-Sumner outlook. Protesting that he had no intention of making moral judgments, Veblen claimed that the simple aggression characteristic of barbarian culture had given way to " shrewd practise and chicanery, as the best approved method of accumulating wealth." These are the qualities which have become essential for selective admission into the leisure class. " The tendency of the pecuniary life is, in a general way, to conserve the barbarian temperament, but with the substitution of fraud and prudence, or administrative ability, in place of the predilection for physical damage that characterizes the early barbarian." The process of selection, under the conditions of modern society, has caused the aristocratic and bourgeois virtues — " that is to say the destructive and pecuniary traits " — to be found among the upper classes, and the industrial virtues, the peaceable traits, largely among " the classes given to mechanical industry." [24]

A more fundamental aspect of Veblen's use of Darwinism was his criticism of the methods of economic theory, best expressed in an essay entitled " Why Is Economics Not an Evolutionary Science? " which was published in the *Quarterly Journal of Economics* in 1898. What, asked Veblen, is the thing that distinguishes post-Darwinian science from pre-evolutionary science? It is not the insistence on facts, nor again the effort to formulate schemes of growth or development. It is a difference of " spiritual point of view . . . a difference in the basis of valuation of the facts for the scientific purpose, or in the interest from which the facts are appreciated." Evolutionary science is " unwilling to depart from the test of causal relation or quantitative sequence." The modern scientist who asks the question " Why? " demands an answer in terms of cause and effect and refuses to go beyond it to any ultimate system, to any teleological conception of the cosmos. This is the crux of the distinction; for earlier natural scientists were not satisfied with this bare formula of mechanical sequence, but sought for some ultimate systematization of the facts within a framework of " natural law." They persistently clung to the notion of some " spiritually legitimate end " resident in and underlying the matters of fact which they observed. Their object was " to formulate knowledge in terms of absolute truth; and this absolute truth is a spiritual fact."

This pre-Darwinian viewpoint, Veblen insisted, and not that of evolutionary science, still dominates the conceptions of modern economics. The " ultimate laws " formulated by the classical economists are laws setting down the normal or natural in the light of their preconception " regarding the ends to which, in the nature of things, all things tend "; and this preconception " imputes to things a tendency to work out what the instructed common sense of the time accepts as the adequate or worthy end of human effort." Yet evolutionary natural science deals only with cumulative causation, and not with the formulation of some "normal case" which is constructed not out of any available facts but out of the investigator's ideal of economic life. Traditional economics, following a preconceived notion of the normal, formulates an abstraction of the hedonistic man as "a homogeneous globule of desire of happiness," passive under the buffetings of pain and pleasure stimuli. In the light of evolutionary science, on the contrary, man is seen to be "a coherent structure of propensities and habits which seeks realization and expression in an unfolding activity." Instead of seeking for normal cases in the existence of an imaginary normal hedonistic man, a truly evolutionary economics must be "the theory of a process of cultural growth as determined by the economic interest, a theory of cumulative sequence of economic institutions stated in terms of the process itself." [25]

Where other economists had found in Darwinian science merely a source of plausible analogies or a fresh rhetoric to substantiate traditional postulates and precepts, Veblen saw it as a loom upon which the whole fabric of economic thinking could be rewoven. The dominant school of economists had said that the existing is the normal and the normal is the right, and that the roots of human ills lie in acts which interfere with the natural unfolding of this normal process toward its inherent end in a beneficent order.

By virtue of their hedonistic preconceptions, their habituation to the ways of a pecuniary culture, and their avowed animistic faith that nature is in the right, the classical economists knew that the consummation to which, in the nature of things, all things tend, is the frictionless and beneficent economic system. This competitive ideal, therefore, affords the normal, and conformity to its requirements affords the test of absolute economic truth.[26]

In so far as economists had tried to use Darwinism, it was only to fortify this theoretical structure. Henceforth economics should abandon such preconceived notions and devote itself to a theory of the evolution of institutions as they actually are.

Although coinciding with the general atmosphere of protest, Veblen's criticisms were often misunderstood and took effect but slowly. For some time his work was most popular among radicals for whom he had little respect. Yet a quarter-century after his essays on the evolutionary method in economic science, a colleague found him an effective force " with the large group of his disciples and the larger group of those who have been driven to rethink their premises and reorient their efforts by the challenge of his pitiless subversions of orthodoxy." [27]

III

The methods and concepts of sociology, which was still striving for an established place in American schools, underwent a transformation much more sweeping than that in economics. Between 1890 and 1915 sociology was affected by changes both in the social scene and in other disciplines, particularly psychology. So rapidly did sociology develop, and so profuse was its literature, that it is impossible to give a full account of the fate of Darwinian sociology or do more than indicate leading tendencies in theory.

Outstanding sociologists followed either the Spencer-Sumner pattern or the one set by Lester Ward. Ward himself held an increasingly prominent place after 1893, and his election as first president of the American Sociological Society in 1906 was an acknowledgment of his leadership in the field. E. A. Ross and Albion Small considered themselves his disciples. Small, who was particularly concerned with the history and methodology of social science, took a special interest in promoting Ward's works. Ross, who had married Ward's niece, was an enthusiastic follower.

Among Spencerians the leader was still Sumner, who, however, had turned from his individualist sermons to enter upon the massive study that produced *Folkways* and the posthumous *Science of Society*. Sumner's chief disciple, Albert Galloway Keller, extended his master's work by applying in temperate fashion the Darwinian concepts of variation, selection, transmission, and adaptation to human folkways. While Keller's approach was institutional rather than atomistic, reflecting the later phase of Sumner's own development, he was as skeptical as his teacher of proposals for a quick or drastic reconstruction of society, and as heartily devoted to a rigidly deterministic view of social evolution. The persistence of this view emerged most clearly in his attitude toward adaptation. " If we can accept the conclusion . . . that every established and settled institution is justifiable, in its setting, as an adaptation," Keller wrote, " it seems to me that we are thereby accepting the extension of the Darwinian theory to the field of the science of society." [28]

Franklin H. Giddings at Columbia University continued to work with the Spencerian concepts of differentiation and equilibration and similar cosmic principles long after they had been abandoned by most other writers; [29] but he readily admitted that sociology is a psychological rather than a biological science, and was quick to point out that the cornerstone of his own social theory, " the consciousness of kind," which he held to be the basis of all social organization, is a mental state and not a biological process.[30] A thoroughgoing individualist, he took a conservative view of the selection principle in society. Although he recognized that the fittest is not always the best, he held it to be characteristic of the social process to make them identical. Society, however, in selecting the best, gives weight to such qualities as sympathy and mutual aid. It usually eliminates " the incompetent and the irresponsible." [31] Since the days when political economy had been his primary interest, Giddings had been devoted to the competitive principle, and he believed with Spencer that its permanence in the economic process could be deduced from the conservation of energy and the facts of heredity.[32] He drew upon biology to support the ancient doctrine of a natural aristocracy, and accordingly argued for a modification of pure democracy.[33]

The most important change in sociological method was its estrangement from biology, and the tendency to place social studies on a psychological foundation. Spencer had not long completed his *Principles of Sociology* when the tide began to turn powerfully against him, and repudiations of his method were so drastic as to be unmindful of the qualifications Spencer himself had made. Albion Small wrote in 1897:

> Mr. Herbert Spencer has been a much mixed blessing. . . . He has probably done more than any man of recent times to set a fashion of semi-learned thought, but he has lived to hear himself pronounced an anachronism by men who were once his disciples. . . . Mr. Spencer's sociology is of the past, not of the present. . . . Spencer's principles of sociology are supposed principles of biology extended to cover social relations. But the decisive factors in social relations are understood by present sociologists to be psychical, not biological.[34]

Small's attitude was the prevailing one. " I believe that the biologic bias creates erroneous notions of social phenomena, and stimulates activity along fruitless lines of investigation," Simon Patten had declared.[35] Even the Spencerian Giddings was moved to concede that " the attempt to construct a science of society by means of biological analogies has been abandoned by all serious investigators of social phenomena." [36] Ross's *Foundations of Sociology* (1905) contained a critique of Spencerianism and related tendencies. Albion Small observed that the general line of methodological progress in sociology was " marked by a gradual shifting of effort from analogical representation of social structures to real analysis of social processes." [37] Small's own *Introduction to the Study of Society* (1894), written in collaboration with George E. Vincent, had employed a moderate use of analogical representation. Twenty years later Small confessed that " the emptiness of this sort of work now makes my teeth chatter . . ." [38] Charles A. Ellwood, writing fifty years after *The Origin of Species*, found Darwinism of great use to sociology, but disparaged Spencer's attempt to carry over physical and mechanical principles to society. Spencer's interpretations, " being fundamentally in terms alien to the social life, were foredoomed to failure." [39] James Mark Baldwin agreed:

> The attempt . . . very current at one time through the influence of Spencer, to interpret social organization by strict analogy with the physical organism is now discredited. Such a view will not stand before the consideration of the most elementary psychological principles.[40]

The tendency to draw upon psychology instead of biology was in line with Ward's appeals for a proper evaluation of the psychic factor in civilization, and took place under his leadership, but the psychology that the most fruitful innovators in social theory were adopting was less traditional than either Ward's or Spencer's, deriving its impetus largely from the work of James and Dewey. Before their work, psychology had been bound to traditional hedonism. The Spencerian and Wardian conception of human motivation, like that of the classical economists whom Veblen criticized, was substantially bounded by the pleasure-pain, stimulus-response horizon. The new psychology, whose most eminent representatives were Dewey and Veblen portrayed the organism as a structure of propensities, interests, and habits, not as a mere machine for the reception and registering of pleasure-pain stimuli.

The new psychology, moreover, was a truly social psychology. The social conditioning of the individual's reaction pattern was stressed by Dewey and Veblen, and insistence upon the unreality of a personal psyche isolated from the social surroundings was a central tenet in the social theory of Charles H. Cooley and the psychology of James Mark Baldwin.[41] The older psychology had been atomistic; Spencer, for example, had seen society as the more or less automatic result of the characters and instincts of its individual members; and this had given color to his conclusion that the improvement of society must be a slow evolutionary process waiting upon the gradual increment of personal characteristics " adapted " to the life conditions of modern industrial society. The new psychology, prepared to see the interdependence of the individual personality with the institutional structure of society, was destroying this one-way notion of social causation and criticizing its underlying individualism. " The individual," wrote Baldwin, " is a product of his social life and society is an organization of such individuals." [42] The thesis of Cooley's social psychology was " that man's psychical outfit is not divisible into the social and the nonsocial; but that he is all social in a larger sense, is all a part of the common human life . . ." [43] John Dewey analyzed the implications for social action of this view of human nature:

> We may desire abolition of war, industrial justice, greater equality of opportunity for all. But no amount of preaching good will or the golden rule or cultivation of sentiments of love and equity will accomplish the results. There must be change in objective arrangements and institutions. We must work on the environment, not merely on the hearts of men. [44]

The tempo of the change in social theory should not be exaggerated, however. For many years after its appearance in 1908, William McDougall's *Introduction to Social Psychology* was the most popular book in its field; and McDougall was the proponent par excellence of the "fixed structure of mind." McDougall derived the salient features of the human equipment from instincts traceable far back into the biological past of the race. For many who were influenced by McDougall's instinct theory, progress toward a cultural analysis of social phenomena was as difficult as for those who had listened to Spencer a generation before. [45]

Influenced by the humanitarianism of the day and the political renaissance of the common man, the new sociology was pulled along in the current of Progressivism. The discipline was no longer seen by its practitioners as a complicated way of justifying laissez faire. Men like Ross and Cooley refused to look upon the poor as unfit or to worship at the shrine of the fittest. [46] It is significant that Ross, the most popular spokesman of sociology, [47] fits the pattern of the typical Progressive thinker. A midwesterner, a supporter of Populism in his early days, later a friend of many of the muckrakers, he expressed in his formal writings the aggressive spirit of protest and reform. "Suckled on the practicalism of Lester F. Ward," he explained, "I wouldn't give a snap of my finger for the 'pussyfooting' sociologist." [48] In his early work Ross tore down the analogy between natural selection and the economic process, and condemned it as "a caricature of Darwinism, invented to justify the ruthless practices of business men." [49] In his *Sin and Society* (1907) Ross criticized the prevailing code of morals for failing to pierce the veil of the impersonal corporate relations of modern society and to fix blame for social ills on absentee malefactors. The spirit of reform had been set loose within the very discipline that Spencer had hoped would teach men to let things alone.

IV

During these same years when social Darwinism was under increasingly strong criticism among social theorists, it was being revived in a somewhat new guise in the literature of the eugenics movement. Accompanied by a flood of valuable genetic research carried on by physicians and biologists, eugenics seemed not so much a social philosophy as a science; but in the minds of most of its advocates it had serious consequences for social thought.

The theory of natural selection, which had assumed the transmission of parental variations, had greatly stimulated the study of heredity. Popular credulity about the scope and variety of hereditary traits had been almost boundless. Darwin's cousin, Francis Galton, had laid the foundations of the eugenics movement and coined its name during the years when Darwinism was being sold to the public. In the United States, Richard Dugdale had published in 1877 his study of *The Jukes,* which, although its author gave more credit to environmental factors than did many later eugenists,[50] had nevertheless offered support to the common view that disease, pauperism, and immorality are largely controlled by inheritance. While Galton's first inquiries into heredity — *Hereditary Genius* (1869), *Inquiries into Human Faculty* (1883), and *National Inheritance* (1889) — had been received here with much acclaim, it was not until the turn of the century that the eugenics movement took organized form, first in England and then in the United States. Eugenics then grew with such great rapidity that by 1915 it had reached the dimensions of a fad. While eugenics has never since been so widely discussed, it has proved to be the most enduring aspect of social Darwinism.

In 1894, Amos G. Warner, in his standard study of *American Charities*, had wrestled with the problem of the relative importance of heredity and environment in the background of poverty.[51] At the turn of the century there was a notable rise of interest in the social significance of hereditary characteristics.[52] The American Breeders' Association, founded in 1903, rapidly developed a strong eugenics subsection, which by 1913 became influential enough to have the name of the organization changed to the American Genetic Association. In 1910 a group of eugenists, with the financial assistance of Mrs. E. H. Harriman, founded at Cold Spring Harbor the Eugenics Record Office, which became a laboratory and a fountainhead of propaganda.

The National Conference on Race Betterment in 1914 showed how thoroughly the eugenic ideal had made its way into the medical profession, the colleges, social work, and charitable organizations.[53] The ideas of the movement began to receive practical application in 1907, when Indiana became the first state to adopt a sterilization law; by 1915 twelve states had passed similar measures.[54]

Doubtless the rapid urbanization of American life, which created great slums in which were massed the diseased, the deficient, and the demented, had much to do with the rise of eugenics. The movement was also favored by a growing interest in philanthropy and increasing endowments for hospitals and charities and appropriations for public health. Especially stimulating to the study of mental disease and deficiency was the rapid expansion of American psychiatry after 1900. As more and more diseased and defective families in great cities came to the attention of physicians and social workers, it was easy to confuse the rising mass of known cases with a real increase. The influx of a large immigrant population from peasant countries of central and southern Europe, hard to assimilate because of rustic habits and language barriers, gave color to the notion that immigration was lowering the standard of American intelligence; at least so it seemed to nativists who assumed that a glib command of English is a natural criterion of intellectual capacity. The apparent economic deceleration at the end of the century was also seen by many observers as the beginning of a national decline; and it was in accordance with the habits of a Darwinized era to find in this apparent social decline a biological deterioration associated with the disappearance of "the American type." [55]

Among scientists and physicians the movement was spurred by several biological discoveries. Weismann's germ-plasm theory stimulated a hereditarian approach to social theory.[56] The rediscovery in 1900 by DeVries, and others, of Mendel's studies in heredity placed in the hands of geneticists the organizing principle which their inquiries had lacked and gave them fresh confidence in the possibilities of their research for prediction and control.

Few of the eugenists presumed to be social philosophers or to offer a full program of social reconstruction; and they were sometimes careful to qualify their hereditarian theses with a nod to the uses of environment; but this did not prevent them from adopting the biological approach to social analysis at the very time that it was being dropped by leaders in social theory. William E. Kellicott probably spoke for a majority when he said that "the Eugenist believes that no other single factor in determining social conditions and practices approaches in importance that of racial structural integrity and sanity." [57]

Early eugenists tacitly accepted that identification of the "fit" with the upper classes and the "unfit" with the lower that had been characteristic of the older social Darwinism. Their warnings about the multiplication of morons at the lower end of the social scale, and their habit of speaking of the "fit" as if they were all native, well-to-do, college-trained citizens, sustained the old belief that the poor are held down by biological deficiency instead of environmental conditions. Their almost exclusive focus upon the physical and medical aspects of human life helped to distract public attention from the broad problems of social welfare. They were also in large part responsible for the emphasis upon preserving the "racial stock" as a means of national salvation — an emphasis so congenial to militant nationalists like Theodore Roosevelt.[58] They differed, however, from earlier social Darwinists in that they failed to draw sweeping laissez-faire conclusions; indeed a part of their own program depended upon state action. Still, they were almost equally conservative in their general bias; and so authoritative did their biological data seem that they were convincing to men like E. A. Ross, who had thoroughly repudiated Spencerian individualism.

The social preconceptions of Sir Francis Galton were not seriously questioned by the early eugenists; and Galton, like Bowen, Sumner, and Arthur Latham Perry, had postulated the free competitive order in which awards are distributed according to ability. He was convinced " that the men who achieve eminence, and those who are naturally capable, are, to a large extent, identical." " If a man is gifted with vast intellectual ability, eagerness to work, and power of working," he added, " I cannot comprehend how such a man should be repressed." Galton insisted that " social hindrances " cannot prevent men of high ability from becoming eminent, and, on the other hand, that " social advantages are incompetent to give that status to a man of moderate ability." [59]

Karl Pearson set the tone of eugenics on this point when he estimated that heredity accounts for nine-tenths of a man's capacity.[60] Henry Goddard, as a result of his investigation of the Kallikaks, concluded that feeble-mindedness is " largely responsible " for paupers as well as criminals, prostitutes, and drunkards.[61] David Starr Jordan declared that " poverty, dirt and crime " could be ascribed to poor human material, and added, " It is not the strength of the strong but the weakness of the weak which engenders exploitation and tyranny." [62] Lewellys F. Barker, a distinguished physician, suggested that the decline and fall of nations could be explained through the relative fertility of the fit and unfit elements.[63] Charles B. Davenport, the leader of American eugenics, challenged the environmentalist assumptions that dominated current social practice, and argued that " the greatest need of the day for the progress of social science is additional precise data as to the unit characteristics of man and their methods of inheritance." [64]

Edward Lee Thorndike did much to spread among educators the eugenists' idea of inherited mental capacity. Thorndike believed that men's *absolute* achievements can be affected by environment and training, but that their *relative* achievements, their comparative performances in rivalry with each other, can be accounted for only by original capacity.[65] Fundamentally it is the soundness and rationality of the racial stock that creates the environment and not vice versa. "There is no so certain and economical way to improve man's environment as to improve his nature." [66] For educational policy this view demanded the development of the intellectual faculties of the few who have outstanding abilities, and giving limited vocational training to the mediocre.[67]

The consequences for social policy of the eugenic point of view were treated at some length by Popenoe and Johnson in their popular textbook, *Applied Eugenics.* Among the reforms they favored were large inheritance taxes, the back-to-the farm movement, the abolition of child labor, and compulsory education. Rural living would counteract the dysgenic effect of urban society. The abolition of child labor would cause the poor to restrict their breeding. Compulsory education would have the same effect by making the child an expense to its parents; but it should not be supplemented by subsidies to children of the poor in the form of free lunches, free textbooks, or other aids that would lower the cost of child care. The authors opposed minimum-wage legislation and trade unions on the ground that both favored inferior workmen and penalized the superior by fixing wages in industry without regard to individual merit. They also opposed socialism for its belief in the benefits that would flow from environmental changes and for its faith in human equality; but they did break with individualism in so far as eugenics sought a social end requiring some individual subordination.[68]

Although eugenists were given to attacking the Jeffersonian

doctrine of natural equality, few were willing to go far enough to challenge the ideal of democratic government. When Alleyne Ireland, a noted critic of democracy, wrote in the *Journal of Heredity* that Weismann's germ-plasm theory sapped the intellectual foundations of democracy by ruling out the possibility that the inferior could be improved from generation to generation by education and training, he was immediately challenged by biologists who saw no inevitable contradiction between natural inequality and democratic government.[69]

Some biologists had remarkable confidence in their ability to resolve the problems of politics by the methods of science. When the First World War threw the menace of " Kaiserism " into the limelight, Frederick Adams Woods, a student of heredity in royal families, pointed out that the most despotic Roman emperors had been closely related. If despots are largely the result of hereditary forces, he concluded, " then the only way to eliminate despots is to regulate the sources from which they spring." In so far as the despots are recast in their ancestral mold, " the number of despots can be reduced by a control of the marriages from which they originate." [70]

The ideology of the movement drew fire from representatives of the trend toward cultural analysis in sociology. Lester Ward, who had long before tried to refute Galton, saw in the eugenics ideology a menace to his own theories, and he devoted the greater part of his *Applied Sociology* to an attack upon the hereditarian argument. Analyzing the very cases used by Galton to prove that genius is hereditary, Ward showed that opportunity and education were also universally present.[71]

In 1897 Charles H. Cooley, influenced by Ward's own early work,[72] published a critical review of Galton's thesis, pointing out that all his cases of " hereditary genius " had been provided with certain simple tools — literacy and access to books — without which no amount of genius could make its way. Remarking that there had been a very high percentage of illiteracy among the common people of England in the middle of the nineteenth century, Cooley asked how the geniuses in this mass of illiterates could have risen to fame, no matter how great their native endowment.[73] Albert Galloway Keller also reminded eugenists that their proposals involved a thoroughgoing transformation in the mores, above all in the strong and deep-rooted mores of sex.[74] It was Cooley who summarized most pointedly the objections of mature sociologists to the eugenists' conception of social causation:

> Most of the writers on eugenics have been biologists or physicians who have never acquired that point of view which sees in society a psychological organism with a life process of its own. They have thought of human heredity as a tendency to definite modes of conduct, and of environment as something that may aid or hinder, not remembering what they might have learned even from Darwin, that heredity takes on a distinctively human character only by renouncing, as it were, the function of predetermined adaptation and becoming plastic to the environment.[75]

V

In spite of its fundamental conservatism, the eugenics craze had about it the air of a " reform," for it emerged at a time when most Americans liked to think of themselves as reformers. Like the reform movements, eugenics accepted the principle of state action toward a common end and spoke in terms of the collective destiny of the group rather than of individual success.

This is significant of the general trend of thought in the Progressive era. A rising regard for the collective aspects of life was one of the outstanding characteristics of the shift in the dominant pattern of thought. The new collectivism was not socialistic, but was based upon an increasing recognition of the psychological and moral relatedness of men in society. It saw in the coexistence of baronial splendor and grinding poverty something more than the accidental dispensations of Providence. Refusing to depend upon individual self-assertion as an adequate remedy, men turned toward collective social action.

The change in the political outlook of the common man

was responsible for a change in the fundamental mechanisms of thought among workers in the social sciences. The formalistic thought of the nineteenth century had been built upon an atomistic individualism. Society, men had believed, was a loose collection of individual agents; social advance depended upon improvements in the personal qualities of these individuals, their increased energy and frugality; among these individuals the strongest and best rose to the top and gave leadership to the rest; their heroic accomplishments were the ideal subject matter of history; the best laws were those that gave them the greatest scope for their activities; the best nations were those that produced most leaders of this type; the way of salvation was to leave unhindered the natural processes that produced these leaders and gave the affairs of the world into their hands.

This pattern of thought was static; instead of inquiry it seemed to encourage deductive speculation; its essential function was the rationalization of existing institutions. Those who were satisfied with it had felt relatively little need for concrete investigation or even for significant novelty in their abstractions.

Between the Spanish-American War and the outbreak of the First World War there was a great restlessness in American society, which inevitably affected the patterns of speculative thought. The old scheme of thought was repeatedly assailed by critics who were in sympathy with the new spirit of the Progressive era. The intellectual friction engendered by this discontent fired the energies and released the critical talents of new minds in history, economics, sociology, anthropology, and law. The result was a minor renaissance in American social thought, a renaissance which saw in a relatively short span of years the rise to prominence of Charles A. Beard, Frederick Jackson Turner, Thorstein Veblen, John R. Commons, John Dewey, Franz Boas, Louis D. Brandeis, and Oliver Wendell Holmes.

It is easier to enumerate the achievements of this renaissance than to characterize its intellectual assumptions, but certainly its leading figures did share a common consciousness of society as a collective whole rather than a congeries of individual atoms. They shared also an understanding of the need for empirical research and accurate description rather than theoretical speculation cast in some traditional mold.

A drastic departure from ancestor worship in history was marked by Charles Beard's study of the origins of the Constitution and by Frederick Jackson Turner's quest for environmental and economic explanations of American development. Brandeis opened up new possibilities in law by drafting for the first time a factual sociological brief in defense of a state law regulating conditions of labor in private enterprise. Franz Boas led a generation of anthropologists away from unilinear evolutionary theory toward cultural history and took pioneer steps in the criticism of race theory. John Dewey made philosophy a working instrument in other disciplines, applying it fruitfully to psychology, sociology, education, and politics. Veblen exposed the intellectual sterility of prevailing economic theory, and pointed the way to an institutional analysis of the facts of economic life.

In accordance with the spirit of the times, the most original thinkers in social science had ceased to make their main aim the justification and perpetuation of existing society in all its details. They were trying to describe it with accuracy, to understand it in new terms, and to improve it.

Chapter Nine

Racism and Imperialism

The brutality of all national development is apparent, and we make no excuse for it. To conceal it would be a denial of fact; to glamour it over, an apology to truth. There is little in life that is not brutal except our ideal. As we increase the aggregate of individuals and their collective activities, we increase proportionately their brutality.

GENERAL HOMER LEA

In this world the nation that has trained itself to a career of unwarlike and isolated ease is bound, in the end, to go down before other nations which have not lost the manly and adventurous qualities.

THEODORE ROOSEVELT

In 1898 the United States waged a three-month war with Spain. It took the Philippine Islands from Spain by treaty and formally annexed the Hawaiian Islands. In 1899 the United States partitioned the Samoan Islands by agreement with Germany, and expressed its policy toward western interests in China in the " Open Door " note. In 1900 Americans took part in suppressing the Chinese Boxer Rebellion. By 1902 the Army had finally suppressed insurrection in the Philippines; and in that year the islands were made an unorganized territory.

As the United States stepped upon the stage of empire, American thought turned once again to the subjects of war and empire; opponents and defenders of expansion and conquest marshaled arguments for their causes. After the fashion of late nineteenth-century thought, they sought in the world of nature a larger justification for their ideals.

The use of natural selection as a vindication of militarism or imperialism was not new in European or American thought. Imperialists, calling upon Darwinism in defense of the subjugation of weaker races, could point to *The Origin of Species,* which had referred in its subtitle to *The Preservation of Favored Races in the Struggle for Life.* Darwin had been talking about pigeons, but the imperialists saw no reason why his theories should not apply to men, and the whole spirit of the naturalistic world-view seemed to call for a vigorous and unrelenting thoroughness in the application of biological concepts. Had not Darwin himself written complacently in *The Descent of Man* of the likelihood that backward races would disappear before the advance of higher civilizations? [1]   Militarists could also point to the harsh fact of the elimination of the unfit as an urgent reason for cultivating the martial virtues and keeping the national powder dry. After the Franco-Prussian War both sides had for the first time invoked Darwinism as an explanation of the facts of battle.[2]   " The greatest authority of all the advocates of war is Darwin," explained Max Nordau in the *North American Review* in 1889. " Since the theory of evolution has been promulgated, they can cover their natural barbarism with the name of Darwin and proclaim the sanguinary instincts of their inmost hearts as the last word of science." [3]

It would nevertheless be easy to exaggerate the significance of Darwin for race theory or militarism either in the United States or in western Europe. Neither the philosophy of force nor doctrines of *Machtpolitik* had to wait upon Darwin to make their appearance. Nor was racism strictly a post-Darwinian phenomenon. Gobineau's *Essai sur l'Inégalité des Races Humaines,* a landmark in the history of Aryanism, was published in 1853–55 without benefit of the idea of natural selection. As for the United States, a people long familiar with Indian warfare on the frontier and the pro-slavery arguments of Southern politicians and publicists had been thoroughly grounded in notions of racial superiority. At the time when Darwin was still hesitantly outlining his theory in private, racial destiny had already been called upon by American expansionists to support the conquest of Mexico. " The Mexican race now see in the fate of the aborigines of the north, their own inevitably destiny," an expansionist had written. "They must amalgamate or be lost in the superior vigor of the Anglo-Saxon race, or they must utterly perish." [4]

This Anglo-Saxon dogma became the chief element in American racism in the imperial era; but the *mystique* of Anglo-Saxonism, which for a time had a particularly powerful grip on American historians, did not depend upon Darwinism either for its inception or for its development. It is doubtful that such monuments of English Anglo-Saxon historical writing as Edward Augustus Freeman's *History of the Norman Conquest of England* (1867–79) or Charles Kingsley's *The Roman and the Teuton* (1864) owed much to biology; and certainly John Mitchell Kemble's *The Saxons in England* (1849) was not inspired by the survival of the fittest. Like other varieties of racism, Anglo-Saxonism was a product of modern nationalism and the romantic movement rather than an outgrowth of biological science. Even the idea that a nation is an organism that must either grow or fall into decay, which doubtless received an additional impetus from Darwinism, had been invoked before 1859 by the proponents of "Manifest Destiny." [5]

Still, Darwinism was put in the service of the imperial urge. Although Darwinism was not the primary source of the belligerent ideology and dogmatic racism of the late nineteenth century, it did become a new instrument in the hands of the theorists of race and struggle. The likeness of the Darwinian portrait of nature as a field of battle to the prevailing conceptions of a militant age in which von Moltke could write that " war is an element of the order of the world established by God . . . [without which] the world would stagnate and lose itself in materialism," was too great to escape attention. In the United States, however, such frank and brutal militarism was far less common than a benevolent conception of Anglo-Saxon world domination in the interests of peace and freedom. In the decades after 1885, Anglo-Saxonism, belligerent or pacific, was the dominant abstract rationale of American imperialism.

The Darwinian mood sustained the belief in Anglo-Saxon racial superiority which obsessed many American thinkers in the latter half of the nineteenth century. The measure of world dominion already achieved by the " race " seemed to prove it the fittest. Also, in the 1870's and 1880's many of the historical conceptions of the Anglo-Saxon school began to reflect advances in biology and allied developments in other fields of thought. For a time American historians fell under the spell of the scientific ideal and dreamed of evolving a science of history comparable to the biological sciences.[6] The keynote of their faith could be found in E. A. Freeman's *Comparative Politics* (1874), in which he allied the comparative method with the idea of Anglo-Saxon superiority. " For the purposes of the study of Comparative Politics," he had written, " a political constitution is a specimen to be studied, classified, and labeled, as a building or an animal is studied, classified, and labeled by those to whom buildings or animals are objects of study." [7]

If political constitutions were to be classified and compared by Victorian scholars as if they were animal forms, it was highly probable that the political methods of certain peoples would be favored over others. Inspired by the results of the comparative method in philology and mythology, particularly by the work of Edward Tylor and Max Müller, Freeman tried, using this method, to trace the signs of original unity in the primitive institutions of the Aryans, particularly in the " three most illustrious branches of the common stock — the Greek, the Roman, and the Teuton."

When Herbert Baxter Adams set up his great historical seminar at Johns Hopkins, it was with the official blessing of Freeman; and Freeman's dictum, " History is past politics and politics is present history," was emblazoned on the historical studies that came pouring forth from Adams' seminar. A whole generation of historians receiving their inspiration from the Johns Hopkins school could have said with Henry Adams, " I flung myself obediently into the arms of the Anglo-Saxons in history." [8] The leading notion of the Anglo-Saxon school was that the democratic institutions of England and the United States, particularly the New England town meeting could be traced back to the primitive institutions of the early German tribes.⁹ In spite of differences in detail, the Hopkins historians were in general agreement on their picture of the big, blond, democratic Teuton and on the Teutonic genealogy of self-government. The viewpoint of the school was given a fitting popular expression in 1890 with the publication of James K. Hosmer's *Short History of Anglo-Saxon Freedom*, which drew upon the whole literature of Anglo-Saxondom to establish the thesis that government of the people and by the people is of ancient Anglo-Saxon origin. Wrote Hosmer:

> Though Anglo-Saxon freedom in a more or less partial form has been adopted (it would be better perhaps to say imitated) by every nation in Europe, but Russia, and in Asia by Japan, the hopes for that freedom, in the future, rest with the English-speaking race. By that race alone it has been preserved amidst a thousand perils; to that race alone is it thoroughly congenial; if we can conceive the possibility of the disappearance among peoples of that race, the chance would be small for that freedom's survival . . .¹⁰

Hosmer shared the optimism of his English contemporary John Richard Green, who believed that the English-speaking race would grow in enormous numbers and spread over the New World, Africa, and Australia. " The inevitable issue," concluded Hosmer, " is to be that the primacy of the world will lie with us. English institutions, English speech, English thought, are to become the main features of the political, social, and intellectual life of mankind." ¹¹ Thus would the survival of the fittest be written large in the world's political future.

What Hosmer did for Anglo-Saxon history, John W. Burgess did for political theory. His *Political Science and Comparative Constitutional Law*, published in the same year as Hosmer's book, serves as a reminder of German as well as English influences in the American Anglo-Saxon cult; for Burgess, like Herbert Baxter Adams, had received a large part of his graduate training in Germany. The peculiarity of his work, Burgess declared, was its method. " It is a comparative study. It is an attempt to apply the method, which has been found so productive in the domain of Natural Science, to Political Science and Jurisprudence." It was Burgess' contention that political capacity is not a gift common to all nations, but limited to a few. The highest capacity for political organization, he believed, has been shown, in unequal degrees, by the Aryan nations. Of all these, only " the Teuton really dominates the world by his superior political genius."

It is therefore not to be assumed that every nation must become a state. The political subjection or attachment of unpolitical nations to those possessing political endowment appears, if we may judge from history, to be as truly a part of the world's civilization as is the national organization of states. I do not think that Asia and Africa can ever receive political organization in any other way. . . . The national state is . . . the most modern and complete solution of the whole problem of political organization which the world has yet produced; and the fact that it is the creation of Teutonic political genius stamps the Teutonic nations as the political nations *par excellence,* and authorizes them, in the economy of the world, to assume the leadership in the establishment and administration of states. . . . The Teutonic nations can never regard the exercise of political power as a right of man. With them this power must be based upon capacity to discharge political duty, and they themselves are the best organs which have as yet appeared to determine when and where this capacity exists.[12]

Theodore Roosevelt, who had been Burgess' student at Columbia Law School was also inspired by the drama of racial expansion. In his historical work, *The Winning of the West,* Roosevelt drew from the story of the frontiersman's struggle with the Indians the conclusion that the coming of the whites was not to be stayed and a racial war to the finish was inevitable.[13] "During the past three centuries," wrote the young scholar-in-politics, "the spread of the English-speaking peoples over the world's waste spaces has been not only the most striking feature in the world's history, but also the event of all others most far-reaching in its effects and its importance." This great expansion he traced back many centuries to the days when German tribes went forth to conquest from their marshy forests. American development represents the culminating achievement of this mighty history of racial growth.[14]

The writings of John Fiske, one of the earliest American synthesizers of evolutionism, expansionism, and the Anglo-Saxon myth, show how tenuous could be the boundary between Spencer's ideal evolutionary pacifism and the militant imperialism which succeeded it. A kindly man, whose thought was grounded in Spencer's theory of the transition from militancy to industrialism, Fiske was not the sort to advocate violence as an instrument of national policy. Yet even in his hands evolutionary dogma issued forth in a bumptious doctrine of racial destiny. In his *Outlines of Cosmic Philosophy*, Fiske had followed Spencer in accepting the universality of conflict (outside of family relationships) as a fact in savage society; he believed it an effective agent in selection.[15] But the superior, more differentiated and integrated societies had come to prevail over the more backward by natural selection, and the power of making war on a grand scale had become concentrated in the hands of " those communities in which predatory activity is at the minimum and industrial activity at the maximum." So warfare or destructive competition gives place to the productive competition of industrial society.[16] As militancy declines, the method of conquest is replaced by the method of federation.

Fiske, who had long believed in Aryan race superiority,[17] also accepted the " Teutonic " theory of democracy.[18] This doctrine sanctified any conquest incidental to Anglo-Saxon expansion. English victories over France in the eighteenth-century colonial struggles represented a victory for industrialism over militancy. The American victory over Spain and the acquisition of the Philippines Fiske interpreted as the high point in a conflict between Spanish colonization and superior English methods.[19]

In 1880, when he was invited to speak before the Royal Institute of Great Britain, Fiske gave a series of three lectures on " American Political Ideas " which became widely known as a statement of the Anglo-Saxon thesis. Fiske praised the ancient Roman Empire as an agency of peace, but argued that it had been inadequate as a system of political organization because it failed to combine concerted action with local self-government. The solution to this ancient need could be provided by representative democracy and the local self-government embodied in the New England town. By retaining the rustic democracy of America's Aryan forefathers, American federal organization would make possible an effective union of many diverse states. Democracy, diversity, and peace would be brought into harmony. The dispersion of this magnificent Aryan political system over the world, and the complete elimination of warfare, was the next step in world history.

With characteristic Darwinian emphasis upon race fertility, Fiske dwelt upon the great population potential of the English and American races. America could support at least 700,000,000; and the English people would within a few centuries cover Africa with teeming cities, flourishing farms, railroads, telegraphs, and all the devices of civilization. This was the Manifest Destiny of the race. Every land on the globe that was not already the seat of an old civilization should become English in language, traditions, and blood. Four-fifths of the human race would trace its pedigree to English forefathers. Spread from the rising to the setting sun, the race would hold the sovereignty of the sea and the commercial supremacy which it had begun to acquire when England first began to settle the New World.[20] If the United States would only drop its shameful tariff and enter into free competition with the rest of the world, it would exert such pressure, peacefully of course, that the states of Europe would no longer be able to afford armaments and would finally see the advantages of peace and federation. Thus, according to Fiske, would man finally pass out of barbarism and become truly Christian.[21]

Even Fiske, who was accustomed to platform success, was astonished at the enthusiasm evoked by these addresses in England and at home.[22] The lecture on "Manifest Destiny," published in *Harper's* in 1885, was repeated more than twenty times in cities throughout the United States.[23] By request of President Hayes, Chief Justice Waite, Senators Hoar and Dawes of Massachusetts, General Sherman, George

Bancroft, and others, Fiske gave his lectures again at Washington, where he was feted by the politicos and presented to the Cabinet.[24]

As a spokesman of expansion, however, Fiske was but a small voice compared with the Rev. Josiah Strong, whose book *Our Country: Its Possible Future and Its Present Crisis*, appeared in 1885 and soon sold 175,000 copies in English alone. Strong, then secretary of the Evangelical Society of the United States, wrote the book primarily to solicit money for missions. His uncanny capacity for assimilating the writings of Darwin and Spencer to the prejudices of rural Protestant America makes the book one of the most revealing documents of its time. Strong exulted in the material resources of the United States, but he was dissatisfied with its spiritual life. He was against immigrants, Catholics, Mormons, saloons, tobacco, large cities, socialists, and concentrated wealth — all grave menaces to the Republic. Still he was undaunted in his faith in universal progress, material and moral, and the future of the Anglo-Saxon race. He employed the economic argument for imperialism; and a decade before Frederick Jackson Turner he saw in the imminent exhaustion of the public lands a turning point in national development. It was Anglo-Saxonism, however, that brought him to the highest pitch of enthusiasm. The Anglo-Saxon people, the bearers of civil liberty and pure spiritual Christianity, said Strong,

. . . is multiplying more rapidly than any other European race. It already owns one-third of the earth, and will get more as it grows. By 1980 the world Anglo-Saxon race should number at least 713,000,000. Since North America is much bigger than the little English isle, it will be the seat of Anglo-Saxondom.

If human progress follows a law of development, if " Time's noblest offspring is the last," our civilization should be the noblest; for we are " The heirs of all the ages in the foremost files of time," and not only do we occupy the latitude of power, but *our land is the last to be occupied in that latitude.* There is no other virgin soil in the North Temperate Zone. If the consummation of human progress is not to be looked for here, if there is yet to flower a higher civilization, where is the soil that is to produce it?[25]

Strong went on to show how a new and finer physical type was emerging in the United States, bigger, stronger, taller than Scots or Englishmen. Darwin himself, Strong noted triumphantly, had seen in the superior vigor of Americans an illustration of natural selection at work, when he wrote in *The Descent of Man*:

There is apparently much truth in the belief that the wonderful progress of the United States, as well as the character of the people, are the results of natural selection; for the more energetic, restless, and courageous men from all parts of Europe have emigrated during the last ten or twelve generations to that great country, and have there succeeded best. Looking to the distant future, I do not think that the Reverend Mr. Zincke takes an exaggerated view when he says: "All other series of events — as that which resulted in the culture of mind in Greece, and that which resulted in the empire of Rome — only appear to have purpose and value when viewed in connection with, or rather as subsidiary to . . . the great stream of Anglo-Saxon emigration to the west." [26]

Returning to his theme that the unoccupied lands of the world were filling up, and that population would soon be pressing upon subsistence in the United States as in Europe and Asia, Strong declared:

Then will the world enter upon a new stage of its history — *the final competition of races for which the Anglo-Saxon is being schooled.* If I do not read amiss, this powerful race will move down upon Mexico, down upon Central and South America, out upon the islands of the sea, over upon Africa and beyond. And can anyone doubt that the result of this competition of races will be the "survival of the fittest"? [27]

II

Although concrete economic and strategic interests, such as Chinese trade and the vital necessity of sea power, were the prominent issues in the imperial debate, the movement took its rationale from more general ideological conceptions. The appeal of Anglo-Saxonism was reflected in the adherence to it of political leaders of the expansion movement. The idea of inevitable Anglo-Saxon destiny figured in the outlook of Senators Albert T. Beveridge and Henry Cabot Lodge and of John Hay, Theodore Roosevelt's Secretary of State, as

well as the President himself. During the fight for the annexation of the Philippines, when the larger question of imperial policy was thrown open for debate, expansionists were
quick to invoke the law of progress, the inevitable tendency
to expand, the Manifest Destiny of Anglo-Saxons, and the
survival of the fittest. Before the Senate in 1899, Beveridge
cried:

> God has not been preparing the English-speaking and Teutonic peo-
> ples for a thousand years for nothing but vain and idle self-admiration.
> No! He has made us the master organizers of the world to establish
> system where chaos reigns. . . . He has made us adepts in govern-
> ment that we may administer government among savages and senile
> peoples.[28]

In the most memorable of his imperialist exhortations,
"The Strenuous Life" (1899), Theodore Roosevelt warned
of the possibility of national elimination in the international
struggle for existence:

> We cannot avoid the responsibilities that confront us in Hawaii,
> Cuba, Porto Rico, and the Philippines. All we can decide is whether
> we shall meet them in a way that will redound to the national credit,
> or whether we shall make of our dealings with these new problems a
> dark and shameful page in our history. . . . The timid man, the
> lazy man, the man who distrusts his country, the over-civilized man,
> who has lost the great fighting, masterful virtues, the ignorant man, and
> the man of dull mind, whose soul is incapable of feeling the mighty lift
> that thrills "stern men with empires in their brains" — all these, of
> course, shrink from seeing the nation undertake its new duties. . . .
> I preach to you, then, my countrymen, that our country calls not
> for the life of ease but for the life of strenuous endeavor. The twenti-
> eth century looms before us big with the fate of many nations. If we
> stand idly by, if we seek merely swollen, slothful ease and ignoble peace,
> if we shrink from the hard contests where men must win at hazard of
> their lives and at the risk of all they hold dear, then the bolder and
> stronger peoples will pass us by, and will win for themselves the domi-
> nation of the world.[29]

John Hay found in the impulse to expand a sign of an
irresistible "cosmic tendency." "No man, no party, can
fight with any chance of final success against a cosmic tendency; no cleverness, no popularity avails against the spirit of

the age." [30] " If history teaches any lesson," echoed another writer a few years later, " it is that nations, like individuals, follow the law of their being; that in their growth and in their decline they are creatures of conditions in which their own volition plays but a part, and that often the smallest part." [31] The question of the Philippines was sometimes pictured as the watershed of American destiny; our decision would determine whether we should undergo a new expansion greater than any in the past, or fall back into decline as a senile people. Said John Barrett, former minister to Siam:

Now is the critical time when the United States should strain every nerve and bend all her energies to keep well in front in the mighty struggle that has begun for the supremacy of the Pacific Seas. If we seize the opportunity we may become leaders forever, but if we are laggards now we will remain laggards until the crack of doom. The rule of the survival of the fittest applies to nations as well as to the animal kingdom. It is a cruel, relentless principle being exercised in a cruel, relentless competition of mighty forces; and these will trample over us without sympathy or remorse unless we are trained to endure and strong enough to stand the pace.[32]

Charles A. Conant, a prominent journalist and economist troubled about the necessity of finding an outlet for surplus capital, " if the entire fabric of the present economic order is not to be shaken by a social revolution," argued that

. . . the law of self-preservation, as well as that of the survival of the fittest, is urging our people on in a path which is undoubtedly a departure from the policy of the past, but which is inevitably marked out by the new conditions and requirements of the present.[33]

Conant warned against the possibility of decadence if the country did not seize upon its opportunities at once.[34] Another writer denied that a policy of colonial expansion was anything novel in American history. We had colonized the West. The question was not whether we should now enter upon a colonial career but whether we should shift our colonizing heritage into new channels. " We must not forget that the Anglo-Saxon race is expansive." [35]

Although the Anglo-Saxon *mystique* was called upon in the

interests of expansion by might, it also had its more pacific side. Its devotees had usually recognized a powerful bond with England; the historians of the Anglo-Saxon school, stressing the common political heritage, wrote about the American Revolution as if it were a temporary misunderstanding in a long history of common political evolution, or a welcome stimulant to flagging Anglo-Saxon liberties.

One outgrowth of the Anglo-Saxon legend was a movement toward an Anglo-American alliance which came to rapid fruition in the closing years of the nineteenth century. In spite of its unflagging conviction of racial superiority, this movement was peaceful rather than militaristic in its motivation; for its followers generally believed that an Anglo-American understanding, alliance, or federation would usher in a " golden age " of universal peace and freedom.[36] No possible power or combination of powers would be strong enough to challenge such a union. This " English-speaking people's league of God for the permanent peace of this war-worn world," as Senator Beveridge called it, would be the next stage in the world's evolution. Advocates of Anglo-American unity believed that Spencer's transition from militant to pacific culture, and Tennyson's " Parliament of Man, the Federation of the World," were about to become a reality.

James K. Hosmer had appealed in 1890 for an " English-Speaking Fraternity " powerful enough to withstand any challenge by the Slavs, Hindus, or Chinese. This coalescence of like-minded states would be but the first step toward a brotherhood of humanity.[37] Yet it was not until 1897 that American interest in an English alliance resulted in a movement of consequence, which received the support of publicists and statesmen as well as littérateurs and historians. During the war with Spain, when continental nations took a predominantly hostile attitude toward American interests, Britain's friendliness stood out in welcome relief. Common fears of Russia and a feeling of identity of interests in the Far East were added to the notion of a common racial destiny. The Anglophobia which had been so persistent among American politicians — Roosevelt and Lodge had been among the bitterest — was considerably relieved. The anti-imperialist Carl Schurz felt that what he rather prematurely took to be the complete dissipation of anti-English feeling was one of the best results of the Spanish-American War.[38] Richard Olney — who as Cleveland's Secretary of State during the Venezuela dispute had defiantly told Britain that the fiat of the United States is law in the Western Hemisphere — now wrote an article on " The International Isolation of the United States " to point out the benefits of British trade and to warn against pursuing an anti-British policy at a time when our country stood alone in the world.[39] Arguing that " family quarrels " were a thing of the past, Olney expressed his hope for Anglo-American diplomatic coöperation, and reminded his readers: " There is a patriotism of race as well as of country." Even the navalist Mahan approved of the British, and although he had felt for some time that a movement for union was premature, he was sufficiently friendly to be content to let the British retain naval supremacy.[40] For a short time at the close of the century the Anglo-Saxon movement became the rage among the upper classes, and statesmen spoke seriously of a possible political alliance.[41]

The Anglo-Saxon cult, however, had to pull against the great mass of the population, whose ethnic composition and cultural background rendered them immune to its propaganda; and even among those of Anglo-Saxon lineage the dynamic appeal of the cult was confined to the years of excitement at the turn of the century. The term " Anglo-Saxon " offended many people, and meetings of protest against Anglo-Saxonism were called in some of the western states.[42] Suspicion of England, traditional in American politics, could not be overcome. John Hay complained in 1900 of " a mad-dog hatred of England prevalent among newspapers and politicians." [43] When the movement for Anglo-American Union was revived again during the First World War, the term " English-speaking " was used in preference to " Anglo-

Saxon," and racial exclusiveness was no longer featured.[44] The powerful undertow of American isolation that followed the war, however, swept away this movement once again.

Anglo-Saxonism in politics was limited both in scope and in duration. It had its day of influence as a doctrine of national self-assertion, but as a doctrine of Anglo-Saxon world order its effects were ephemeral. Even the benevolent ideal of the dreamers of a Pax Anglo-Americana found practical meaning only as a timely justification of a temporary rapprochement inspired by the needs of *Realpolitik*. The day had not come when world peace could be imposed by a " superior " race confident in its biological blessings and its divine mission.

III

Lacking an influential military caste, the United States never developed a strong military cult audacious enough to glorify war for its own sake. Such outbursts as Roosevelt's " Strenuous Life " speech were rare; and it was also rare for an American writer to extol war for its effects upon the race, although Rear Admiral Stephen B. Luce, one of Mahan's patrons, once declared that war is one of the great agencies of human conflict and that " strife in one form or another in the organic world seems to be the law of existence. . . . Suspend the struggle, well called the battle of life, for a brief space, and death claims the victory." [45] Most writers on war seemed to agree with Spencer that military conflict had been highly useful in developing primitive civilization but had now long outlived its value as an instrument of progress.[46]

The advocates of preparedness did not usually take the stand that there is anything inherently desirable in war, but rather quoted the old maxim, " If you wish for peace, prepare for war." " Let us worship peace, indeed," conceded Mahan, " as the goal at which humanity must hope to arrive; but let us not fancy that peace is to be had as a boy wrenches an unripe fruit from a tree." [47]

Others took the position that strife is inherent in the nature of things and must be anticipated as an unhappy necessity. Once the martial fever of the short and easy war with Spain had subsided, the psychology of the American people between 1898 and 1917 was surprisingly nervous and defensive for a nation that was rapidly rising in stature as a world power. Encouraged by the eugenics movement, men talked of racial degeneracy, of race suicide, of the decline of western civilization, of the effeteness of the western peoples, of the Yellow Peril. Warnings of decay were most commonly coupled with exhortations to revivify the national spirit.

One of the most popular among the pessimistic writers was an Englishman, Charles Pearson, who had formerly served the Empire as minister of education in Victoria. His melancholy book, *National Life and Character*, published in England and the United States in 1893, offered a discouraging prognosis for western culture. The higher races, Pearson believed, can live only in the temperate zone, and will be forever barred from effective colonization in the tropics. Overpopulation and economic exigencies will give rise to state socialism, which will extend its tentacles into every corner of western national life. Because of the increasing dependence of the citizen upon the state, nationalism will grow, and religion, family life, and old-fashioned morality will decline. There will also be a consolidation of peoples into great centralized empires, for only these will have the capacity to survive. Large armies, great cities, huge national debts will hasten cultural eclipse. The decline of competition, coupled with state education, will render the intellect more mechanical in its operations and deprive it of the initiative that alone is capable of outstanding achievement in the arts. The result will be a world of old people, scientific rather than esthetic, unprogressive, stable, without adventure, energy, brightness, hope, or ambition. Meanwhile other races will not fail in vitality, for biology shows that the lower are more prolific than the higher. Chinese, Hindus, Negroes cannot be exterminated, but will on the contrary be likely to challenge the supremacy of western civilization by industrial rather than military means. Perhaps the best that the governing races can do is to face the future with courage and dignity.

It is idle to say that if all this should come to pass our pride of place will not be humiliated. We were struggling among ourselves for supremacy in a world which we thought of as destined to belong to the Aryan races and to the Christian faith; to the letters and arts and charm of social manners which we have inherited from the best times of the past. We shall wake to find ourselves elbowed and hustled, and perhaps even thrust aside by peoples whom we looked down upon as servile, and thought of as bound always to minister to our needs. The solitary consolation will be that the changes have been inevitable. It has been our work to organize and create, to carry peace and law and order over the world, that others may enter in and enjoy. Yet in some of us the feeling of caste is so strong that we are not sorry to think we shall have passed away before that day arrives.[48]

Pearson's fears were the beginning of a reaction from the optimism expressed by Fiske and Strong in the 1880's. For middle-class intellectuals, reeling under the shock of the panic of 1893 and the deep social discontents of the prolonged depression that followed, his prophecies of doom had a ring of truth. They were particularly suited to the dark mood that overcame Henry Adams in the 1890's. He wrote to C. M. Gaskell:

I am satisfied that Pearson is right, and that the dark races are gaining on us, as they have already done in Haiti, and are doing throughout the West Indies and our Southern States. In another fifty years, at the same rate of movement, the white races will have to reconquer the tropics by war and nomadic invasion, or be shut up, north of the fortieth parallel.[49]

To his brother, Brooks Adams, pessimism was more than a matter of private despair. In his study of *The Law of Civilization and Decay* (1896), he set forth his own version of the deeper historical principles behind the façade of social change. The law of force and energy is universal, said Adams in a passage somewhat reminiscent of Spencer, and animal life is only one of the outlets through which solar energy is dissipated. Human societies are forms of animal life, differing in energy according to their natural endowments; but all societies obey the general law that the social movement of a community is proportionate to its energy and mass, and that its degree of centralization is proportionate to its mass. The surplus energetic material not expended by a society in the daily struggle for life can be stored as wealth, and the stored energy is transmitted from one community to another either by conquest or by superiority in economic competition. Every race sooner or later reaches the limit of its warlike energy and enters upon a phase of economic competition. Surplus energy, when accumulated in such bulk as to preponderate over productive energy, becomes the controlling social force. Capital becomes autocratic. The economic and scientific intellect grows at the expense of imaginative, emotional, and martial arts. A stationary period may supervene, lasting until it is terminated by war or exhaustion or both.

The evidence, however, seems to point to the conclusion that, when a highly centralized society disintegrates, under the pressure of economic competition, it is because the energy of the race has been exhausted. Consequently, the survivors of such a community lack the power necessary for renewed concentration, and must probably remain inert until supplied with fresh energetic material by the infusion of barbarian blood.[60]

In subsequent volumes, *America's Economic Supremacy* (1900) and *The New Empire* (1902), Adams worked out a materialistic interpretation of society based upon physics, biology, geography, and economics. Surveying the rise and decline of historic states, he attributed changes in supremacy to changes in basic trade routes. The center of economic civilization, now once again in transit, he saw coming to rest in the United States; but he warned that "supremacy has always entailed its sacrifices as well as its triumphs, and fortune has seldom smiled on those who, beside being energetic and industrious, have not been armed, organized, and bold."[61]

Nature tends to favor organisms that operate most cheaply — that is, with the most economic expenditure of energy. Wasteful organisms are rejected by nature; they can be eliminated by commerce if not by conquest. Adams was particularly anxious about a possible conflict with Russia in the east, for which he thought the United States should be well armed.[52] Concerning the tendency toward centralized empires, he wrote:

> Moreover, Americans must recognize that this is war to the death, — a struggle no longer against single nations but against a continent. There is not room in the economy of the world for two centres of wealth and empire. One organism, in the end, will destroy the other. The weaker must succumb. Under commercial competition, that society will survive which works cheapest; but to be undersold is often more fatal to a population than to be conquered.[53]

More influential than Brooks Adams was Captain Alfred Thayer Mahan, whose book *The Influence of Sea Power upon History* (1890) had made him the world's most prominent exponent of navalism. In *The Interest of America in Sea Power* (1897), in which he urged that the country pursue a stronger policy than the present one of " passive self-defense," Mahan pointed out:

> All around us now is strife; "the struggle of life," "the race of life," are phrases so familiar that we do not feel their significance till we stop to think about them. Everywhere nation is arrayed against nation; our own no less than others.[54]

Theodore Roosevelt was among those who tried to stir the nation against the eventualities predicted by Pearson and foreseen by Brooks Adams. For Pearson's pessimism he saw little excuse; although he conceded that civilized nations were not destined to rule the tropics, he could not believe that the white races would lose heart or become intimidated by the tropic races. When western institutions, and democratic government itself, spread to the tropics, the danger of an overpowering industrial competition would be considerably less; and it seemed unlikely that high industrial efficiency would be achieved without a marked degree of westernization. He was somewhat more favorably impressed with the work of his friend Brooks Adams, but again the most pessimistic prophecies aroused Roosevelt to reply. He did not believe that the martial type of man necessarily decays as civilization progresses; pointing to the examples of Russia and Spain, he argued that the phenomenon of national decline should not be too closely identified with advancing industrialism. Only when Adams mentioned the failure to produce enough healthy children did he touch upon the real danger to our society.[55] This was a theme dear to Roosevelt's heart. Vociferously fearful of the menace of race decadence through decline in the birth rate, he never tired of the theme of reproduction and motherhood. If marriages did not produce an average of four children, the numbers of the race could not be maintained. He warned that if the process of racial decay continued in the United States and the British Empire, the future of the white race would rest in the hands of the German and the Slav.[56]

Associated with fears of racial decline and of the loss of fighting fiber was the menace of the Yellow Peril, which was much talked about between 1905 and 1916.[57] The prevailing western attitude toward Japan had been friendly until the Japanese victory over Russia in 1905. However, with the convincing demonstration of the Japanese martial prowess, attitudes changed, just as they had toward Germany after her victory in 1871.[58] In the United States, fear of the Japanese was especially strong in California, where oriental immigration had been resented for over thirty years.[59] The sensational press took up the Japanese menace and exploited it to the point of stimulating occasional war scares.[60]

In 1904 Jack London, always a strenuous advocate of racial assertiveness, warned in an article in the *San Francisco Examiner* of the potential threat to the Anglo-Saxon world if the organizing and ruling capacities of the Japanese should ever gain control of the enormous working capacity of the great Chinese population. The impending racial conflict, he thought, might come to a head in his own time.

The possibility of race adventure has not passed away. We are in the midst of our own. The Slav is just girding himself up to begin. Why may not the yellow and brown start out on an adventure as tremendous as our own and more strikingly unique? [61]

Hugh H. Lusk believed that the Japanese menace was only a small part of a general reawakening of the Mongolian race.

whose urge to expansion, motivated by the age-old population problem, might soon send it out over the Pacific and ultimately to southwestern America and to the gates of the United States via Mexico.[62] Talk of the Yellow Peril reached its height just before the First World War, when congressmen spoke openly of inevitable conflict in the Pacific.[63]

Perhaps the closest American approximation to the German militarist writer General von Bernhardi was General Homer Lea, a colorful military adventurer who fought against the Boxer Rebellion, and later became an adviser to Sun Yat-sen. Lea's militarism was based directly upon biology. He believed that nations are like organisms in their dependence upon growth and expansion to resist disease and decay.

As physical vigor represents the strength of man in his struggle for existence, in the same sense military vigor constitutes the strength of nations; ideals, laws and constitutions are but temporary effulgences, and are existent only so long as this strength remains vital. As manhood marks the height of physical vigor among mankind, so the militant successes of a nation mark the zenith of its physical greatness.[64]

Militancy may be divided into three phases: the militancy of the struggle to survive, the militancy of conquest, and the militancy of supremacy or preservation of ownership. It is in the first stage, the struggle to survive, that the genius of a people reaches its height; the harder this struggle, the more highly developed is the military spirit, with the result that conquerors often arise from desolate wastes or rocky islands. The laws of struggle and survival are universal and unalterable, and the duration of national existence is dependent upon the knowledge of them.

Plans to thwart them, to short-cut them, circumvent, to cozen, to scorn and violate [them] is folly such as man's conceit alone makes possible. Never has this been tried — and man is ever at it — but what the end has been gangrenous and fatal.[65]

Lea warned of the possibility of Japanese invasion of the United States, and argued that a war with Japan would be settled by land campaigns, for which the country needed a much larger army. Without such a military establishment, the West Coast would stand in deadly danger of invasion. The strategy of such an invasion Lea had planned in full detail.

Lea further warned that the Saxon races were flouting the laws of nature by permitting the militancy of their people to decline. A decadent tendency to let individual wants take precedence over the necessities of national existence threatened Anglo-Saxon power throughout the world, he believed. The United States, submerged by a flood of non-Anglo-Saxon immigrants, was ceasing to be the stronghold of a Saxon race. The British Empire was in serious danger from the colored races. The day of the Saxon was ending. For the impending struggle between the Germans and the Saxon race, the latter was ill equipped. There was only one antidote for Anglo-Saxon decline: greater militancy. A confederation would be weak in war, but universal compulsory military service might check the already alarming decline.[66]

The advocates of preparedness made a biological appeal similar to Lea's. Hudson Maxim, an inventor of smokeless powder, and brother of Hiram Maxim, the inventor of the Maxim gun, published a volume called *Defenseless America* (1914), which was widely distributed by Hearst's International Library. " Self-preservation," Maxim warned, " is the first law of Nature, and this law applies to nations exactly as it applies to individuals. Our American Republic cannot survive unless it obeys the law of survival." He argued that man is by nature a struggling animal, that human nature has always been more or less the same. To be unprepared for the struggle would be to risk extinction, but preparedness might avert war.[67]

A similar philosophy could be found among the wartime leaders of the organized preparedness movement.[68] S. Stanwood Menken, chairman of the National Security League's Congress of Constructive Patriotism, warned the delegates that the law of the survival of the fittest applied to nations, and that the United States could assert its fitness only through a national reawakening.[69] General Leonard Wood was skep-

tical of the possibility of suppressing war, which, he said,
" is about as difficult as to effectively neutralize the general
law which governs all things, namely the survival of the fittest." [70]   Although the biological argument for militarism
was hardly the dominant note among American leaders, it
did give them a cosmic foundation that appealed to a Darwinized national mentality.

IV

In 1898, when the problem of expansion had arisen, the
anti-imperialists had not been inclined to answer the racial
appeal or to dislocate it from its Darwinian framework.
They preferred to ignore the broad theme of racial destiny,
concentrating instead upon an appeal to American traditions.
The accident of party alignment doubtless had something to
do with the unwillingness of politically minded anti-expansionists to assault the dogma of Anglo-Saxon racial superiority; for the Democratic Party, strongest in the Solid South,
was the bulwark of the opposition, and to deny the Anglo-
Saxon myth would serve only to stir up a race question without answering the fundamental arguments of expansionist
leaders.   What some Democrats did do, however, was to in-
vert the racial aspect of expansion and use it as an argument
against annexation of overseas territories.   The idea was
advanced in Congress, particularly by some of the Southern
members, that to assume the government of the Filipinos
would be to introduce into our political structure an alien,
uncongenial, unassimilable people, probably incapable of
reaching Anglo-Saxon heights in the matter of democratic
self-government.   Senator John W. Daniel of Virginia declared in 1899:

> There is one thing that neither time nor education can change. You
> may change the leopard's spots, but you will never change the differ-
> ent qualities of the races which God has created in order that they
> may fulfill separate and distinct missions in the cultivation and civiliza-
> tion of the world.[71]

Men of scientific training had not yet taken the advanced
position on racial equipotentiality that anthropology now

encourages, and the notion had not been widely popularized. Exceptions there were, of course. In 1894 Franz Boas, in his fresh and skeptical address as vice-president of the Anthropological Section of the American Association for the Advancement of Science, had made a cogent criticism of prevailing attitudes toward the colored races. The unwarranted assumption was commonly made, he pointed out, that because the whites' state of civilization is " higher," their racial aptitudes are higher. The standards of white culture are naïvely posited as a norm, and every deviation from the norm is automatically considered characteristic of a lower type. Boas attributed the cultural superiority of Europeans to the circumstances of their historical development rather than to inherent capacities.[72]

William Z. Ripley's substantial study of *The Races of Europe* (1897) also introduced educated readers to some of the complexities of the idea of race, and discredited the Aryan myth. Among others than specialists or curious laymen, however, there was little understanding of these matters, and for the practical purposes of partisan discussion the complacent assertions of the Anglo-Saxon myth were unanswerable except by appeals to other prejudices. Common among men of learning was the conception, taken over from Haeckel's Biogenetic Law, that, since the development of the individual is a recapitulation of the development of the race, primitives must be considered as being in the arrested stages of childhood or adolescence — " half devil and half child," as Rudyard Kipling had said.[73] This view was accepted by the eminent psychologist and educator G. Stanley Hall in his study of *Adolescence*. Although Hall felt that the childlike character of backward peoples entitled them to tender and sympathetic treatment by their phylogenetic " elders," who should be ashamed to make war on children, the condescending approach to primitive culture underlying the recapitulation theory was not calculated to disturb the spokesmen of racial superiority.[74]

It took a measure of courage, in this climate of opinion, to issue a challenge to the dogma of racial inequality. There

were few who would go so far as Ernest Howard Crosby, an
American disciple of Tolstoi, who wrote of " an Anglo-Saxon
union for the vulgarization of the world," and implied in his
famous parody of Kipling that the benefits of western civilization were not the ideal thing for the slow peoples of outly-
ing islands.[75] However, support came from William James,
who thought we had " destroyed in Luzon the one sacred
thing in the world, the spontaneous budding of a national
life." [76] While few anti-imperialists were ready to challenge
the basic assumption of white or Anglo-Saxon superiority,
there were some who doubted the benefits of spreading civilization by conquest or annexation. These skeptics might
well have agreed with the colored trooper in one of the regi-
ments dispatched to suppress Aguinaldo's rebels in the Philippines, who remarked in a moment of war-weariness, " Dis
shyar white man's burden ain't all it's cracked up to be." [77]

The most usable argument for the anti-imperialists was
to appeal to the traditions of Americanism, a procedure that
introduced no new and unfamiliar ideas. Expansion, it was
argued, would mean the adoption of races alien in language,
customs, and institutions. It would mean the beginnings of
a colonial bureaucracy. It would be aping the way of Britain. It would involve the support of a large standing army,
with a consequent heavy tax burden. To launch upon the
government and exploitation of a helpless people would
shame the finest traditions of American democracy, which
had always insisted upon the legitimacy of government only
with the consent of the governed. A nation so rich and great
within its own continental borders had no pressing need for
further expansion; it would risk much to gain little. Launching upon an imperial career would bring America full square
into the game of world politics, with all its militaristic hatreds and extravagances. Behind this would lurk the constant menace of war for the defense of overseas possessions.[78]

One of the most spirited of the anti-imperialists was William James, who at one time served as vice-president of the
Anti-Imperialist League. From time to time James wrote
indignant letters to the Boston *Evening Transcript* denounc-

ing expansionist ideology. Of the white-man's-burden, manifest-destiny thesis, he complained:

> Could there be a more damning indictment of that whole bloated idolatered "modern civilization" than this amounts to? Civilization is, then, the big, hollow, resounding, corrupting, sophisticating, confusing torrent of mere brutal momentum and irrationality that brings forth fruits like this! [79]

In a counterblast to Roosevelt's speech on the "Strenuous Life," he asserted that Roosevelt was "still mentally in the *Sturm und Drang* period of early adolescence," making speeches about human affairs "from the sole point of view of the organic excitement and difficulty they may bring," and gushing over war as the ideal condition of human society. Of worthwhile ends Roosevelt had "not a word . . . one foe is as good as another, for aught he tells us. . . . He swamps everything together in one flood of abstract bellicose emotion." [80]

William Graham Sumner also attacked the imperial impulse with practically all the weapons in the arsenal of the anti-expansionists. Those who were familiar with Sumner's crisp iconoclasm on the subject of democracy may have rubbed their eyes to see the intransigent schoolmaster attack imperialists for preparing the abandonment of the nation's democratic principles; but his argument had an unquestionable ring of sincerity, particularly since it once again put in jeopardy his position at Yale. "My patriotism," he cried, "is of the kind which is outraged by the notion that the United States never was a great nation until in a petty three months' campaign it knocked to pieces a poor, decrepit, bankrupt old state like Spain." [81]

Probably the best known of all the peace advocates and anti-expansionists was David Starr Jordan, president of Stanford University. More than any other man, Jordan established in the American mind the idea that war is a biological evil rather than a biological blessing, because it carries off the physically and mentally fit and leaves behind the less fit. Jordan, who had lost an elder brother in the Civil War, in

1898 became interested in disarmament and the movement for international arbitration. An eminent biologist and a leader of the eugenics movement, he turned his attention to the biological aspects of war. In a series of volumes published between the Spanish-American War and the First World War, Jordan expounded his thesis, using motley evidence from anthropometrics, casualty statistics, reminiscences of Civil War veterans, and the conclusions of other biologists. Darwin himself, Jordan pointed out, had agreed that war is dysgenic.[82] Jordan became the favorite butt of patriots, militarists, and preparedness advocates, who pointed to continued racial improvement in past eras of constant warfare as evidence against his thesis.[83]

Although Jordan was unsuccessful in imposing his quasipacifistic outlook upon the nation, he did leave a profound conviction of the degenerative effect of war upon the breed; and his doctrine, strengthened by the general reaction against militarism in the years after the First World War, became sanctified by repetition in the most conventional of sources. The editor of the *Saturday Evening Post,* for example, wrote in 1921:

Disarm or die. That is the alternative that confronts all men who dare look. Men who are not afraid to face facts know that just as Nature kills off the weak and unfit, so war wipes out the strong and courageous and robs the race of its most vital blood.[84]

Ironically, the United States entered the First World War in the name not of militarism but of anti-militarism. The consequence was that the wartime climate of opinion was, on the whole, hostile to biological militarism. This, it was felt, was the enemy's philosophy. To intellectuals, the social Darwinism of *Machtpolitik* was an integral part of the philosophy they were fighting against.[85] A feature of the image of brutal German military leadership that emerged from the war literature was the idea that the German mind was dominated by a self-conscious, willful, iron-mailed philosophy of immoralism. The Germans, it was maintained, worshiped Treitschke, Nietzsche, von Bernhardi, and other militarists, who assured them that they were the élite of mankind, a race of supermen destined to conquer Europe or the world, who preached that might makes right, that war is a biological necessity, and that conquest is justified by the survival of the fittest. There was a sudden efflorescence of popular interest in Nietzsche and von Bernhardi. " The name of Nietzsche," commented Paul Elmer More, as early as October 1914, " is beginning by the aid of the daily press to take on a sinister meaning for the man on the street." [86]

British and American scholars who ransacked the literature of German chauvinism did not fail to produce damaging evidence. To " a good war halloweth any cause " and similar effusions from the pen of Nietzsche could be added a long series of damning quotations. " The old churchmen preached of war as a just judgment of God," Klaus Wagner had said in his *Krieg* (1906) ; " the modern natural scientists see in a war a propitious mode of selection." [87] " War," said von Bernhardi in his widely reprinted *Germany and the Next War,*

. . . is not merely a necessary element in the life of nations but an indispensable factor of culture, in which a truly civilized nation finds the highest expression of strength and vitality. .   War gives a biologically just decision, since its decisions rest on the very nature of things. . . . It is not only a biological law, but a moral obligation, and, as such, an indispensable factor in civilization.[88]

The war brought a veritable avalanche of anthologies of similar offensive sayings taken from German philosophers, statesmen, and military leaders. The most scholarly of these, *Conquest and Kultur, Aims of the Germans in Their Own Words,* edited by Wallace Notestein and Elmer E. Stoll, was issued under the auspices of George Creel's Committee on Public Information, and thus received official sanction. The historian and biographer William Roscoe Thayer, who was especially active in propagating this interpretation of the German mentality, declared:

In all directions the Germans saw proof that they were the Chosen People. They interpreted the doctrine of evolution so as to draw from it a warrant for their aspirations. Evolution taught that "the fittest survived."

The champions of the philosophy of supermania lean heavily on biology to support their creed. They have been misled by the phrase "the survival of the fittest." You might infer, to hear them buzz, that only the fittest survive, or, to put it conversely, the fact that you survive is proof that you are the "fittest." [89]

When those who had actually read Nietzsche pointed out that he had nothing but contempt for German chauvinism,[90] it was said that the dominant idea emerging from his acknowledged contradictions was that of German diplomacy and German militarism.[91] Bishop J. Edward Mercer, alarmed at the tendency to show that Nietzsche's thought derived from Darwinism, wrote a defense of Darwin for the English *Nineteenth Century*, playing up Darwin's theory of the moral sense and dissociating him from Nietzsche.[92] The conventional image persisted, however, and was accepted even by scholars who knew Germany well.[93]

The necessity of combatting the philosophy of force led Professor Ralph Barton Perry into a formidable assault upon social Darwinism and all its works. His *Present Conflict of Ideals* (1918) was the most substantial of all the refutations of the Darwinized ethics and sociology that had culminated in the monstrosities attributed to von Bernhardi and Nietzsche.[94] The whole evolutionary dogma, the Darwin-Spencer legacy of progress, the glib optimism of John Fiske, the warnings of Benjamin Kidd, the natural-selection economics of Thomas Nixon Carver — all fell under Professor Perry's axe. Like William James before him, Perry pointed out the essential circularity of the Darwinian sociology, in which power and strength are defined in terms of survival, and survival is in turn explained by strength and power. In the Darwinian view, all changes in types of survival and kinds of fitness are considered without relation to ulterior values; there is no value beyond survival itself. Rome conquering the world by force of arms is as good as Greece conquering it by force of ideas or Judea conquering it by force of religious sentiment. Indeed, because of its biological origins, this view actually shows a "strong tendency to favor the cruder and more violent forms of struggle, as being more unmistakably biological." [95]

Pacifists also took advantage of the reaction from the philosophy of force.[96] At the instance of Norman Angell, George Nasmyth published in 1916 his *Social Progress and the Darwinian Theory*, a popularization of the work of Kropotkin and the Russian sociologist Jacques Novicow, the most eminent continental critic of social Darwinism.[97] "Instead of subjecting it to the searching analysis demanded by its practical social importance," Nasmyth declared, "the intellectual world and public opinion has accepted 'social Darwinism' uncritically and by almost unanimous consent as an integral part of the theory of evolution." For this he believed Spencer was chiefly responsible. The primary biological error of social Darwinism is its habit of ignoring the physical universe, of assuming that the cause of progress is not the struggle of man with his environment but rather the struggle of man with man, which in fact yields nothing. Another error is the misinterpretation of the "fittest" as the strongest or even the most brutal, while to Darwin it meant merely the best adapted to existing conditions. Struggle is also confused with the total death of the vanquished, whereas this selective factor hardly ever operates among men. The entire phenomenon of mutual aid is ignored by the philosophy of force. It is to this that man owes his dominant position in the universe. In a large sense, all mankind is an association, and all wars are civil wars; yet the philosophers of force have never advocated civil war as a source of progress.[98]

> With the exception of a few noteworthy books the subject of sociology is still in a state of complete incoherence. Biological phenomena are confused with social facts. Men who call themselves specialists in the subject can still seriously identify the relations between Germany and France, for example, with those between a cat and a rat without doing great injury to their reputation and without exciting much ridicule.[99]

There were curious by-products of this reaction against militarism. Vernon Kellogg, a biologist who had become ac-

quainted with several German military leaders while serving
under Herbert Hoover in Belgium during the First World
War, reported in a volume on his experiences that the philosophy of the foe was a crude Darwinism ruthlessly applied
to the affairs of nations.[100]   Coming to the attention of William Jennings Bryan, Kellogg's book reinforced his fundamentalist conviction of the inherent evilness of evolutionary
ideas and his determination to wage a crusade against them.[101]
John T. Scopes suffered not only for the theories of Darwin,
but for Wilhelm as well.   For many years Bryan had been
troubled about the possible social implications of Darwinism.   In 1905 E. A. Ross, then teaching at Nebraska University, had found Bryan reading *The Descent of Man,* and
Bryan had told him that such teachings would " weaken the
cause of democracy and strengthen class pride and the power
of wealth." [102]   Here, as in other matters, Bryan had sound
intuitions that his intellect had not the power to discipline.

Chapter Ten

Conclusion

The entire modern deification of survival *per se*, survival returning to itself, survival naked and abstract, with the denial of any substantive excellence in *what* survives, except the capacity for more survival still, is surely the strangest intellectual stopping-place ever proposed by one man to another.

WILLIAM JAMES

There was nothing in Darwinism that inevitably made it an apology for competition or force. Kropotkin's interpretation of Darwinism was as logical as Sumner's. Ward's rejection of biology as a source of social principles was no less natural than Spencer's assumption of a universal dynamic common to biology and society alike. The Christian denial of Darwinian " realism " in social theory was no less natural, as a human reaction, than the harsh logic of the " scientific school." Darwinism had from the first this dual potentiality; intrinsically it was a neutral instrument, capable of supporting opposite ideologies. How, then, can one account for the ascendency, until the 1890's, of the rugged individualist's interpretation of Darwinism?

The answer is that American society saw its own image in the tooth-and-claw version of natural selection, and that its dominant groups were therefore able to dramatize this vision of competition as a thing good in itself. Ruthless business rivalry and unprincipled politics seemed to be justified by the survival philosophy. As long as the dream of personal conquest and individual assertion motivated the middle class, this philosophy seemed tenable, and its critics remained a minority.

This version of Darwinism depended for its continuance upon a general acceptance of unrestrained competition. But nothing is so unstable as " pure " business competition;

nothing is so disastrous to the unlucky or unskilled competitor; nothing, as Benjamin Kidd foresaw, is so difficult as to keep the growing number of the " unfit " reconciled to the operations of such a regime. In time the American middle class shrank from the principle it had glorified, turned in flight from the hideous image of rampant competitive brutality, and repudiated the once heroic entrepreneur as a despoiler of the nation's wealth and morals and a monopolist of its opportunities.

With this reaction came the first conclusive victories of the critics of Darwinian individualism — although it is pertinent to note that the material gains of political and economic reformers were far less complete than their ideological triumphs. When Americans were once in the mood to listen to critics of Darwinian individualism, it was no difficult task for these critics to destroy its flimsy logical structure and persuade their audiences that it had all been a ghastly mistake. Spencer, and the men of Spencer's generation in America, thought that he had written a grand preface to destiny. Their sons came to wonder at its monumental dullness and its quaint self-confidence, and thought of it — if they thought of it at all — only as a revealing commentary on a dead age.

While Darwinian individualism declined, Darwinian collectivism of the nationalist or racist variety was beginning to take hold. Darwinism was made to fit the mold of international conflict-ideologies (a process that had been going on in Europe for a long time) just when its inapplicability to domestic economics was becoming apparent. It had been possible for the theorists of reform to show that, in nature, group cohesion and solidarity had been of value to survival and that individual self-assertion was the exception, not the rule. At a time of imperialist friction there was nothing to stop the advocates of expansion and the propagandists of militarism from invoking these very shibboleths of group survival, or from transmuting them into a doctrine of group assertiveness and racial destiny to justify the ways of international competition. The survival of the fittest had once been used chiefly to support business competition at home; now it was used to support expansion abroad.

These dogmas were employed with success until the outbreak of the First World War. Then, ironically, the " Anglo-Saxon " peoples were swept by a revulsion from international violence. They now turned about and with one voice accused the enemy of being the sole advocate of " racial " aggression and militarism. One-sided and false as it was, the notion that the Germans had a monopoly of militaristic thought had at least the compensation that it put the American people in a frame of mind to repudiate such dogmas. Forever after, Darwinian militarism sounded too much like dangerous German talk.

As a conscious philosophy, social Darwinism had largely disappeared in America by the end of the war. It is significant that since 1914 there has been far less Darwinian individualism in America than there was in the latter decades of the nineteenth century. There were, of course, still at large and in places of responsibility men who thought that Sumner's essays were the last word in economics. Darwinian individualism has persisted as a part of political folklore, even though its rhetoric is seldom heard in formal discussion; the folklore of politics can embrace contradictions that are less admissible in self-conscious social theory. But, with these allowances, it is safe to say that Darwinian individualism is no longer congenial to the mood of the nation.

A resurgence of social Darwinism, in either its individualist or imperialist uses, is always a possibility so long as there is a strong element of predacity in society.[1] Biologists will continue to make technical criticisms of natural selection as a theory of development, but these criticisms are not likely to affect social thought. This is true partly because the phrase "survival of the fittest" has a fixed place in the public mind, and partly because of the complexity and the esoteric quality of technical criticisms.

There is certainly some interaction between social ideas and social institutions. Ideas have effects as well as causes. The history of Darwinian individualism, however, is a clear example of the principle that changes in the structure of social ideas wait on general changes in economic and political life. In determining whether such ideas are accepted, truth and logic are less important criteria than suitability to the intellectual needs and preconceptions of social interests. This is one of the great difficulties that must be faced by rational strategists of social change.

Whatever the course of social philosophy in the future, however, a few conclusions are now accepted by most humanists: that such biological ideas as the " survival of the fittest," whatever their doubtful value in natural science, are utterly useless in attempting to understand society; that the life of man in society, while it is incidentally a biological fact, has characteristics that are not reducible to biology and must be explained in the distinctive terms of a cultural analysis; that the physical well-being of men is a result of their social organization and not vice versa; that social improvement is a product of advances in technology and social organization, not of breeding or selective elimination; that judgments as to the value of competition between men or enterprises or nations must be based upon social and not allegedly biological consequences; and, finally, that there is nothing in nature or a naturalistic philosophy of life to make impossible the acceptance of moral sanctions that can be employed for the common good.

Bibliography

AUTHOR'S NOTE FOR REVISED EDITION: It would be impossible to add references to every relevant book or article that has appeared since the first edition. I have therefore made no effort to expand this bibliography, which itself lists only some selected works that were of special value to me. However, I wish to mention a few books published during the past six years that are of unusual pertinence. Stow Persons, ed., *Evolutionary Thought in America* (New Haven: Yale University Press, 1950), has several valuable essays, which together give a broad survey of the field. Philip P. Wiener, *Evolution and the Founders of Pragmatism* (Cambridge: Harvard University Press, 1949) deals exhaustively with its theme. The transition in American thought dealt with in my later chapters is accounted for at greater length by Morton G. White in his *Social Thought in America* (New York: Viking Press, 1949). The impact of Darwinism upon the whole of American university life and thought is discussed by Walter P. Metzger in Chapter 7 of Richard Hofstadter and Walter P. Metzger, *The Development of Academic Freedom in the United States* (New York: Columbia University Press, 1955).

MANUSCRIPTS

Sumner: The papers of the Sumner Estate available in the Sterling Memorial Library at Yale University do not include Sumner's personal correspondence. The available papers shed light chiefly upon Sumner's pre-academic career.

Ward: The papers of Lester Ward in the John Hay Library at Brown University consists chiefly of thirteen volumes of letters received by Ward. They are of considerable value to one who is interested in estimating Ward's influence. The most revealing of these letters are included in the published collections edited by Bernhard J. Stern, which are cited below. Ward's library has several books with significant annotations, giving unique evidence as to his intellectual interests and opinions.

Students of this phase of American thought are fortunate to have available a vast mass of personal correspondence in printed form. Special collections of letters, and biographies cited below have extensive selections from the correspondences of Charles Darwin.

Herbert Spencer, Asa Gray, John Fiske, Edward Livingston Youmans, Lester Ward, William James, Henry Adams, Theodore Roosevelt, and others.

PERIODICALS

American Journal of Sociology, Chicago, 1896–1920.
Annals of the American Academy of Political and Social Science, Phila·
delphia, 1890–1910.
Appleton's Journal, New York, 1867–81.
Arena, Boston, 1889–99; New York, 1899–1904.
Atlantic Monthly, Boston, 1860–1920.
Forum, New York, 1886–1915.
Galaxy, New York, 1866–78.
Independent, New York, 1860–90.
International Socialist Review, Chicago, 1900–10.
Journal of Heredity, Washington, 1910–19
Journal of Political Economy, Chicago, 1893–1915.
Journal of Speculative Philosophy, St. Louis, 1867–80.
Nation, New York, 1865–1920.
Nationalist, Boston, 1889–91.
North American Review, Boston, 1860–77; New York, 1878–1915.
Popular Science Monthly, New York, 1872–1910.
Psychological Review, Princeton, 1894–1915.

BOOKS

Adams, Brooks. *America's Economic Supremacy.* New York: The
Macmillan Co., 1900.
——. *The Law of Civilization and Decay.* New York: The Macmillan
Co., 1896.
——. *The New Empire.* New York: The Macmillan Co., 1902.
Adams, Henry. *The Education of Henry Adams.* Boston and New
York: Houghton Mifflin Co., 1918.
——. *Letters of Henry Adams* (ed. Worthington C. Ford). Boston:
Houghton Mifflin Co., 1930. 2 vols.

Bagehot, Walter. *Physics and Politics.* New York: D. Appleton & Co.,
1873.
Baldwin, James Mark. *Darwin and the Humanities.* Baltimore: Re·
view Publishing Co., 1909.
——. *The Individual and Society.* Boston: R. G. Badger, 1911.
Barker, Ernest. *Political Thought in England.* New York: Henry
Holt & Co., 1915[?].
Barnes, Harry Elmer, and Becker, Howard. *Contemporary Social
Theory.* New York: D. Appleton-Century Co., 1940.
——. *Social Thought from Lore to Science.* New York: D. C. Heath &
Co., 1938. 2 vols.

Barzun, Jacques. *Darwin, Marx, Wagner.* Boston: Little, Brown & Co., 1941.

Becker, Carl. *The Heavenly City of the Eighteenth Century Philosophers.* New Haven: Yale University Press, 1932.

Behrends, A. J. F. *Socialism and Christianity.* New York: Baker and Taylor, 1886.

Bellamy, Edward. *Edward Bellamy Speaks Again!* Kansas City: The Peerage Press, 1937.

——. *Equality.* New York: D. Appleton & Co., 1897.

——. *Looking Backward.* Boston: Houghton Mifflin Co., 1889.

Boas, Franz. *The Mind of Primitive Man.* New York: The Macmillan Co., 1911.

Brandeis, Louis D. *The Curse of Bigness.* New York: The Viking Press, 1934.

Brinton, Crane. *English Political Thought in the Nineteenth Century.* London: E. Benn, 1933.

Bristol, Lucius M. *Social Adaptation.* Cambridge: Harvard University Press, 1915.

Brooks, Van Wyck. *New England: Indian Summer, 1865–1915.* New York: E. P. Dutton & Co., 1940.

Burgess, John W. *Political Science and Comparative Constitutional Law.* Boston: Ginn & Co., 1890. 2 vols.

Cape, Emily Palmer. *Lester F. Ward, a Personal Sketch.* New York and London: G. P. Putnam's Sons, 1922.

Carver, Thomas Nixon. *Essays in Social Justice.* Cambridge: Harvard University Press, 1915.

——. *The Religion Worth Having.* Boston: Houghton Mifflin Co., 1912.

Chamberlain, Houston Stewart. *The Foundations of the Nineteenth Century.* London: John Lane, 1911. 2 vols.

Chamberlain, John. *Farewell to Reform.* New York: Liveright, 1932.

Clark, John Bates. *The Philosophy of Wealth.* Boston: Ginn & Co., 1885.

Clark, John Spencer. *The Life and Letters of John Fiske.* Boston and New York: Houghton Mifflin Co., 1917. 2 vols.

Cochran, Thomas C., and Miller, William. *The Age of Enterprise.* New York: The Macmillan Co., 1942.

Cooley, Charles Horton. *Human Nature and the Social Order.* New York: Charles Scribner's Sons, 1902.

——. *Social Organization.* New York: Charles Scribner's Sons, 1909.

——. *Social Process.* New York: Charles Scribner's Sons, 1918.

Croly, Herbert. *The Promise of American Life.* New York: The Macmillan Co., 1909.

Curti, Merle E. *Peace or War, the American Struggle, 1636–1936.* New York: W. W. Norton & Co., 1936.

——. *The Social Ideas of American Educators*. New York: Charles Scribner's Sons, 1935.

Darwin, Charles. *The Descent of Man*. London: J. Murray, 1871.
——. *The Origin of Species*. London: J. Murray, 1859.
Darwin, Francis. *The Life and Letters of Charles Darwin*. New York: D. Appleton & Co., 1888. 2 vols.
Davenport, Charles. *Heredity in Relation to Eugenics*. New York: Henry Holt & Co., 1915.
Dewey, John. *Characters and Events* (ed. Joseph Ratner). New York: Henry Holt & Co., 1929. 2 vols.
——. *Democracy and Education*. New York: The Macmillan Co., 1916.
——. *Human Nature and Conduct*. New York: Henry Holt & Co., 1922.
——. *The Influence of Darwin on Philosophy*. New York: Henry Holt & Co., 1910.
——. *The Public and Its Problems*. New York: Henry Holt & Co., 1927.
——, *The Quest for Certainty*. New York: Minton, Balch & Co., 1929.
——. *Reconstruction in Philosophy*. New York: Henry Holt & Co., 1920.
——, and Tufts, James. *Ethics*. New York: Henry Holt & Co., 1908.
Dombrowski, James. *The Early Days of Christian Socialism in America*. New York: Columbia University Press, 1936.
Dorfman, Joseph. *Thorstein Veblen and His America*. New York: The Viking Press, 1934.
Dos Passos, John R. *The Anglo-Saxon Century and the Unification of the English-Speaking People*. New York and London: G. P. Putnam's Sons, 1903.
Drummond, Henry. *The Ascent of Man*. New York: A. L. Burt Co., 1894.
Duncan, David. *The Life and Letters of Herbert Spencer*. New York: D. Appleton & Co., 1908.

Ferri, Enrico. *Socialism and Modern Science*. New York: International Library Publishing Co., 1900.
Fisk, Ethel. *The Letters of John Fiske*. New York: The Macmillan Co., 1940.
Fiske, John. *American Political Ideas*. New York: Harper & Bros., 1885.
——. *A Century of Science and Other Essays*. Boston: Houghton Mifflin & Co., 1899.
——. *Civil Government in the United States*. Boston: Houghton Mifflin & Co., 1890.
——, *The Destiny of Man*. Boston: Houghton Mifflin & Co., 1884.

——. *Edward Livingston Youmans.* New York: D. Appleton & Co., 1894.
——. *Excursions of an Evolutionist.* Boston: Houghton Mifflin & Co., 1884.
——. *The Meaning of Infancy.* Boston: Houghton Mifflin & Co., 1909.
——. *Outlines of Cosmic Philosophy.* Boston: Houghton Mifflin & Co., 1874. 2 vols.

Gabriel, Ralph Henry. *The Course of American Democratic Thought.* New York: The Ronald Press Co., 1940.
Galton, Francis, *Hereditary Genius.* London: Macmillan & Co., 1869.
——. *Inquiries into Human Faculty and Its Development.* London: Macmillan & Co., 1883.
——. *Natural Inheritance.* London and New York: Macmillan & Co., 1889.
Geiger, George R. *The Philosophy of Henry George.* New York: The Macmillan Co., 1933.
George, Henry. *A Perplexed Philosopher.* New York: C. L. Webster & Co., 1892.
——. *Progress and Poverty.* New York, 1879.
——. *Social Problems.* New York: Belford, Clarke, & Co., 1883.
George, Henry, Jr. *The Life of Henry George.* New York: Doubleday and McClure Co., 1900.
Ghent, William J. *Our Benevolent Feudalism.* New York: The Macmillan Co., 1902.
Giddings, Franklin H. *The Elements of Sociology.* New York: The Macmillan Co., 1898.
——. *Inductive Sociology.* New York: The Macmillan Co., 1901.
——. *The Principles of Sociology.* New York: The Macmillan Co., 1896.
——. *The Responsible State.* Boston: Houghton Mifflin Co., 1918.
Gide, Charles, and Rist, Charles. *A History of Economic Doctrines.* Boston: D. C. Heath & Co., 1915.
Gladden, Washington. *Applied Christianity.* Boston: Houghton Mifflin & Co., 1886.
Gobineau, Arthur de. *The Inequality of Human Races* (trans. Adrian Collins). New York: G. P. Putnam's Sons, 1915.
Goldenweiser, Alexander. *History, Psychology, and Culture.* New York: Alfred A. Knopf, 1933.
Grant, Madison. *The Passing of the Great Race.* New York: Charles Scribner's Sons, 1916.
Grattan, C. Hartley. *The Three Jameses.* New York: Longmans, Green & Co., 1932.
Gray, Asa. *Darwiniana.* New York: D. Appleton & Co., 1876.
——. *Letters of Asa Gray* (ed. Jane Loring Gray). Boston: Houghton Mifflin Co., 1893. 2 vols.

Gronlund, Laurence. *The Cooperative Commonwealth.* Boston: Lee and Shepard, 1884.
——. *The New Economy.* New York: H. S. Stone & Co., 1898.
——. *Our Destiny.* Boston: Lee and Shepard, 1890.
Gumplowicz, Ludwig. *The Outlines of Sociology* (trans. Frederick W. Moore). Philadelphia: American Academy of Political and Social Science, 1899.

Haeckel, Ernst. *The Riddle of the Universe.* New York: Harper & Bros., 1900.
Hayes, Carlton J. H. *A Generation of Materialism, 1871–1900.* New York: Harper & Bros., 1941.
Headley, Frederick W. *Darwinism and Modern Socialism.* London: Macmillan & Co., 1909.
Henkin, Leo. *Darwinism in the English Novel.* New York: Corporate Press, 1940.
Hobhouse, Leonard. *Social Evolution and Political Theory.* New York: Columbia University Press, 1911.
——. *Mind in Evolution.* London: Macmillan & Co., 1901.
Hodge, Charles. *What Is Darwinism?* New York: Scribner, Armstrong, & Co., 1874.
Holt, Henry. *Garrulities of an Octogenarian Editor.* Boston: Houghton Mifflin Co., 1923.
Hopkins, Charles Howard. *The Rise of the Social Gospel in American Protestantism, 1865–1915.* New Haven: Yale University Press, 1940.
Huxley, T. H. *Evolution and Ethics and Other Essays.* New York: The Humboldt Publishing Co., 1894.

James, William. *Collected Essays and Reviews.* New York: Longmans, Green & Co., 1920.
——. *The Letters of William James* (ed. Henry James). Boston: The Atlantic Monthly Press, 1920. 2 vols.
——. *Memories and Studies.* New York: Longmans, Green & Co., 1912.
——. *A Pluralistic Universe.* New York: Longmans, Green & Co., 1909.
——. *Pragmatism.* New York: Longmans, Green & Co., 1907.
——. *The Principles of Psychology.* New York: Henry Holt & Co., 1890. 2 vols.
——. *The Will to Believe.* New York: Longmans, Green & Co., 1897.
Josephson, Matthew. *The Politicos.* New York: Harcourt, Brace & Co., 1938.
——. *The President Makers.* New York: Harcourt, Brace & Co., 1940.
——. *The Robber Barons.* New York: Harcourt, Brace & Co., 1934.

Karpf, Fay Berger. *American Social Psychology.* New York and London: McGraw-Hill Book Co., 1932.

Kazin, Alfred. *On Native Grounds*. New York: Reynal & Hitchcock, 1942.

Keller, Albert G. *Reminiscences of William Graham Sumner*. New Haven: Yale University Press, 1933.

Kellicott, William E. *The Social Direction of Human Evolution*. New York: D. Appleton & Co., 1911.

Kellogg, Vernon. *Darwinism To-Day*. New York: Henry Holt & Co., 1907.

Kidd, Benjamin. *Principles of Western Civilization*. New York: The Macmillan Co., 1902.

——. *Social Evolution*. New York: Macmillan & Co., 1894.

Kimball, Elsa P. *Sociology and Education*. New York: Columbia University Press, 1932.

Kraus, Michael. *A History of American History*. New York: Farrar & Rinehart, 1937.

Kropotkin, Peter. *Mutual Aid*. London: W. Heinemann, 1902.

Lea, Homer. *The Day of the Saxon*. New York: Harper & Bros., 1912.

——. *The Valor of Ignorance*. New York: Harper & Bros., 1909.

Lewis, Arthur M. *Evolution, Social and Organic*. Chicago: C. H. Kerr, 1908.

——. *An Introduction to Sociology*. Chicago: C. H. Kerr, 1913.

Lippmann, Walter. *Drift and Mastery*. New York: Mitchell Kennerly, 1914.

Lloyd, Henry Demarest. *Wealth Against Commonwealth*. New York: Harper & Bros., 1894.

London, Charmian. *The Book of Jack London*. New York: The Century Co., 1921. 2 vols.

London, Jack. *Martin Eden*. New York: The Macmillan Co., 1908.

Lowie, Robert H. *The History of Ethnological Theory*. New York: Farrar & Rinehart, 1937.

Lundberg, George A. *et al. Trends in American Sociology*. New York: Harper & Bros., 1929.

McDougall, William. *An Introduction to Social Psychology*. Boston: J. W. Luce & Co., 1909.

Mahan, Alfred Thayer. *The Interest of America in Sea Power*. Boston: Little, Brown & Co., 1897.

Maxim, Hudson. *Defenseless America*. New York: Hearst's International Library Co., 1915.

Nasmyth, George. *Social Progress and the Darwinian Theory*. New York: G. P. Putnam's Sons, 1916.

Nevins, Allan. *The Emergence of Modern America, 1865–1878*. New York: The Macmillan Co., 1928.

Norderskiöld, Erik. *The History of Biology.* New York: Alfred A. Knopf, 1928.

Osborn, Henry Fairfield. *From the Greeks to Darwin.* New York: Charles Scribner's Sons, 1899.

Page, Charles H. *Class and American Sociology.* New York: The Dial Press, 1940.

Parrington, V. L. *Main Currents in American Thought.* New York: Harcourt, Brace & Co., 1927–30. 3 vols.

Patten, Simon. *The Premises of Political Economy.* Philadelphia: J. B. Lippincott Co., 1885.

Pearson, Charles. *National Life and Character.* London and New York: Macmillan & Co., 1893.

Pearson, Karl. *National Life from the Standpoint of Science.* London: A. and C. Black, 1901.

Peirce, Charles Sanders. *Chance, Love, and Logic* (ed. Morris R. Cohen). New York: Harcourt, Brace & Co., 1923.

Perry, Ralph Barton. *Philosophy of the Recent Past.* New York: Charles Scribner's Sons, 1926.

——. *The Present Conflict of Ideals.* New York: Longmans, Green & Co., 1918.

——. *The Thought and Character of William James.* Boston: Little, Brown & Co., 1935. 2 vols.

Popenoe, Paul, and Johnson, Roswell Hill. *Applied Eugenics.* New York: The Macmillan Co., 1918.

Pratt. Julius W. *Expansionists of 1898.* Baltimore: The Johns Hopkins Press, 1936.

Rauschenbusch, Walter. *Christianity and the Social Crisis.* New York: The Macmillan Co., 1907.

——. *Christianizing the Social Order.* New York: The Macmillan Co., 1912.

Riley, Woodbridge. *American Thought from Puritanism to Pragmatism.* New York: Henry Holt & Co., 1915.

Ritchie, David G. *Darwinism and Politics.* London: S. Sonnenschein & Co., 1889.

Rogers, Arthur K. *English and American Philosophy Since 1800.* New York: The Macmillan Co., 1922.

Roosevelt, Theodore. *The New Nationalism.* New York: The Outlook Co., 1910.

——, and Lodge, Henry Cabot. *Selections from the Correspondence of Theodore Roosevelt and Henry Cabot Lodge* (ed. Henry Cabot Lodge). New York: Charles Scribner's Sons, 1925. 2 vols.

——. *The Works of Theodore Roosevelt* (National Ed.). New York: Charles Scribner's Sons, 1926. 20 vols.

Ross, Edward A. *Foundations of Sociology.* New York: The Macmillan Co., 1905.
——. *Seventy Years of It.* New York: D. Appleton-Century Co., 1936.
Rumney, Judah. *Herbert Spencer's Sociology.* London: Williams and Norgate, 1934.

Schilpp, Paul A., ed. *The Philosophy of John Dewey.* Evanston and Chicago: Northwestern University Press, 1939.
Schlesinger, A. M. *The Rise of the City.* New York: The Macmillan Co., 1933.
Schurman, Jacob Gould. *The Ethical Import of Darwinism.* New York: Charles Scribner's Sons, 1887.
Singer, Charles J. *A Short History of Biology.* Oxford: The Clarendon Press, 1931.
Small, Albion W. *General Sociology.* Chicago: The University of Chicago Press, 1905.
——, and Vincent, George E. *An Introduction to the Study of Society.* New York: American Book Co., 1894.
Spencer, Herbert. *An Autobiography.* New York: D. Appleton & Co., 1904. 2 vols.
——. *First Principles.* New York: D. Appleton & Co., 1864.
——. *The Man Versus the State* (ed. Truxton Beale). New York: Mitchell Kennerley, 1916.
——. *The Principles of Ethics.* New York: D. Appleton & Co., 1895–98. 2 vols.
——. *The Principles of Sociology.* New York: D. Appleton & Co., 1876–97. 3 vols.
——. *Social Statics.* New York: D. Appleton & Co., 1864.
——. *The Study of Sociology.* New York: D. Appleton & Co., 1874.
Starr, Harris E. *William Graham Sumner.* New York: Henry Holt & Co., 1925.
Stern, Bernhard J. *Lewis Henry Morgan, Social Evolutionist.* Chicago: University of Chicago Press, 1931.
——, ed. *Young Ward's Diary.* New York: G. P. Putnam's Sons, 1935.
Stoddard, Lothrop. *The Rising Tide of Color.* New York: Charles Scribner's Sons, 1920.
Strong, Josiah. *Our Country.* New York: The American Home Missionary Society, 1885.
Sumner, William G. *The Challenge of Facts and Other Essays.* New Haven: Yale University Press, 1914.
——. *Earth-Hunger and Other Essays.* New Haven: Yale University Press, 1913.
——. *Essays of William Graham Sumner* (ed. Albert G. Keller and Maurice R. Davie). New Haven: Yale University Press, 1934. 2 vols.

——. *Folkways.* Boston: Ginn & Co., 1906.
——. *What Social Classes Owe to Each Other.* New York: Harper & Bros., 1883.
——, and Keller, Albert G. *The Science of Society.* New Haven: Yale University Press, 1927. 4 vols.

Tarbell, Ida. *The Nationalizing of Business, 1878–1898.* New York: The Macmillan Co., 1936.
Thomson, J. Arthur. *Darwinism and Human Life.* New York: The Macmillan Co., 1911.
Townshend, Harvey G. *Philosophical Ideas in the United States.* New York: American Book Co., 1934.

Veblen, Thorstein. *Absentee Ownership.* New York: B. W. Huebsch, 1923.
——. *Essays in Our Changing Order* (ed. Leon Ardzrooni). New York: The Viking Press, 1934.
——. *The Instinct of Workmanship.* New York: B. W. Huebsch, 1914.
——. *The Place of Science in Modern Civilization.* New York: B. W. Huebsch, 1919.
——. *The Theory of Business Enterprise.* New York: Charles Scribner's Sons, 1904.
——. *The Theory of the Leisure Class.* New York: The Macmillan Co., 1899.

Walling, William English. *The Larger Aspects of Socialism.* New York: The Macmillan Co., 1913.
Walker, Francis A. *The Wages Question.* New York: Henry Holt & Co., 1876.
Ward, Lester. *Applied Sociology.* Boston: Ginn & Co., 1906.
——. *Dynamic Sociology.* New York: D. Appleton & Co., 1883. 2 vols.
——. *Glimpses of the Cosmos.* New York: G. P. Putnam's Sons, 1913–18. 6 vols.
——. *Outlines of Sociology.* New York: The Macmillan Co., 1898.
——. *The Psychic Factors of Civilization.* Boston: Ginn & Co., 1893.
——. *Pure Sociology.* New York: The Macmillan Co., 1903.
Warner, Amos G. *American Charities.* New York: T. Y. Crowell & Co., 1894.
Weinberg, Albert K. *Manifest Destiny.* Baltimore: The Johns Hopkins Press, 1935.
Weyl, Walter. *The New Democracy.* New York: The Macmillan Co., 1912.
Wright, Chauncey. *Philosophical Discussions.* New York: Henry Holt & Co., 1877.

Youmans, Edward Livingston, ed. *Herbert Spencer on the Americans and the Americans on Herbert Spencer.* New York: D. Appleton & Co., 1883.

Young, Arthur N. *The Single Tax Movement in the United States.* Princeton: Princeton University Press, 1916.

ARTICLES

Boas, Franz. "Human Faculty as Determined by Race," *Proceedings, American Association for the Advancement of Science,* XLIII (1894), 301–27.

Case, Clarence M. "Eugenics as a Social Philosophy," *Journal of Applied Sociology,* VII (1922), 1–12.

Cochran, Thomas C. "The Faith of Our Fathers," *Frontiers of Democracy,* VI (1939), 17–19.

Cooley, Charles H. "Genius, Fame, and the Comparison of Races," *Annals of the American Academy of Political and Social Science,* IX (1897), 317–58.

Dewey, John. "Evolution and Ethics," *Monist,* VIII (1898), 321–41.

——. "Social Psychology," *Psychological Review,* I (1894), 400–11.

Ely, Richard T. "The Past and the Present of Political Economy," *John Hopkins University Studies in Historical and Political Science,* II (1884).

Harrington, Fred H. "Literary Aspects of American Anti-Imperialism," *New England Quarterly,* X (1937), 650–67.

Hofstadter, Richard. "William Graham Sumner, Social Darwinist," *New England Quarterly,* XIV (1941), 457–77.

Huxley, Thomas Henry. "Administrative Nihilism," *Fortnightly Review,* N. S., XVI (1880), 525–43.

James, William. "Great Men, Great Thoughts, and the Environment," *Atlantic Monthly,* XLVI (1880), 441–59. (Reprinted in *The Will to Believe.*)

Loewnberg, Bert J. "The Reaction of American Scientists to Darwinism," *American Historical Review,* XXXVIII (1933), 687–701.

——. "Darwinism Comes to America," *Mississippi Valley Historical Review,* XXVIII (1941), 339–69.

——. "The Controversy over Evolution in New England, 1859–1873," *New England Quarterly,* VII (1935), 232–57.

Patten, Simon. "The Failure of Biologic Sociology," *Annals of the American Academy of Political and Social Science,* IV (1894), 919–47.

Pratt, Julius W. "The Ideology of American Expansion," in *Essays in Honor of William E. Dodd* (Chicago: University of Chicago Press, 1935), 335–53.

Ratner, Sidney. "Evolution and the Rise of the Scientific Spirit in America," *Philosophy of Science,* III (1936), 104–22.

Saveth, Edward. "Race and Nationalism in American Historiography: The Late Nineteenth Century," *Political Science Quarterly*, LIV (1939), 425–41.

Schlesinger, A. M. "A Critical Period in American Religion, 1875–1900," *Massachusetts Historical Society Proceedings*, LXIV (1932), 523–47.

Small, Albion. "Fifty Years of Sociology in the United States, 1865–1915," *American Journal of Sociology*, XVI (1916), 721–864.

Spencer, Herbert. "A Theory of Population, Deduced from the General Law of Animal Fertility," *Westminster Review*, LVII (1852), 468–501.

Spiller, G[ustav]. "Darwinism and Sociology," *Sociological Review* (Manchester), VII (1914), 232–53.

Stern, Bernhard J. "Giddings, Ward, and Small: An Interchange of Letters," *Social Forces*, X (1932), 305–18.

——, ed. "The Letters of Ludwig Gumplowicz to Lester Ward," *Sociologus* (Leipzig), Beiheft I, 1933.

——, ed. "The Letters of Albion W. Small to Lester F. Ward," *Social Forces*, XII (1933), 163–73; XIII (1935), 323–40; XV (1936), 174–86; XV (1937), 305–27.

——, ed. "Letters of Alfred Russel Wallace to Lester F. Ward," *Scientific Monthly*, XL (1935), 375–79.

——, ed. "The Ward-Ross Correspondence, 1891–1896," *American Sociological Review*, III (1938), 362–401.

Veblen, Thorstein. "Why Is Economics Not an Evolutionary Science?" *Quarterly Journal of Economics*, XIII (1898), 373–97.

Wells, Colin, *et al.*, "Social Darwinism," *American Journal of Sociology*, XII (1907), 695–716.

Notes

In the case of works listed in the bibliography (pages 205–216), the place and date of publication are not repeated here.

CHAPTER 1
THE COMING OF DARWINISM

1. See Francis Darwin, *The Life and Letters of Charles Darwin*, I, 51, 99.
2. On the early activities of Fiske and Youmans, see John Spencer Clark, *Life and Letters of John Fiske*, Vol. I; Ethel Fisk, *The Letters of John Fiske*; John Fiske, *Edward Livingston Youmans*.
3. Henry Fairfield Osborn, *From the Greeks to Darwin*, esp. chap. v.
4. See Arthur M. Schlesinger, "A Critical Period in American Religion, 1875–1900," *Proceedings*, Massachusetts Historical Society, LXIV (1932), 525–27.
5. Clark, *op. cit.*, I, 237.
6. *The Education of Henry Adams* (New York: Modern Library, 1931), pp. 225–26.
7. Emma Brace, *The Life of Charles Loring Brace* (New York, 1894), pp. 300–2.
8. See Bert J. Loewenberg, "The Reaction of American Scientists to Darwinism," *American Historical Review*, XXXVIII (1933), 687.
9. One prominent evolutionist, Edward Drinker Cope, adhered to Lamarckism rather than Darwinism. See H. F. Osborn, *Cope: Master Naturalist* (Princeton, 1931). Many biologists, although convinced by Darwin's data of the validity of the development hypothesis, were critical of his theory of natural selection as an explanation of development. The distinction between the two ideas was not always clearly made in the popular controversies.
10. Agassiz, "Evolution and Permanence of Type," *Atlantic Monthly*, XXXIII (1874), 92–101.
11. C. F. Holder, *Louis Agassiz* (New York, 1893), p. 181; cf. also Agassiz, *op. cit.*, p. 94.
12. Le Conte, *Autobiography* (New York, 1913), p. 287. "It has even been said," acknowledged Agassiz, "that I have myself furnished the strongest evidence of the transmutation theory." Agassiz, *op. cit.*, pp. 100–1.
13. Ralph Barton Perry, *The Thought and Character of William James*, I, 265–66. But see James's later tribute to Agassiz, *Memories and Studies* (New York, 1912).

14. "Scientific Teaching in the Colleges," *Popular Science Monthly*, XVI (1880), 558–59; see also the address of Professor Edward S. Morse, *Proceedings*, American Association for the Advancement of Science, XXV (1876), 140.

15. *Darwiniana*, pp. 9–16; see also Gray's article, "Darwin and His Reviewers," *Atlantic Monthly*, VI (1860), 406–25.

16. Morse, *op. cit.;* Morse's summary included the name of practically every outstanding naturalist in the country, among them E. D. Cope, Joseph Leidy, O. C. Marsh, N. S. Shaler, and Jeffries Wyman.

17. Charles Schuchert and Clara Mae Le Vene, *O. C. Marsh, Pioneer in Paleontology* (New Haven, 1940), p. 247.

18. Charles W. Eliot, "The New Education — Its Organization," *Atlantic Monthly*, XXXIII (1869), 203–20, 358–67.

19. Clark, *op. cit.*, I, 353–76; Fiske, *Outlines of Cosmic Philosophy*, preface, p. vii.

20. Schuchert and Le Vene, *op. cit.*, pp. 238–39.

21. Harris E. Starr, *William Graham Sumner*, pp. 345–69.

22. Henry Holt, *Garrulities of an Octogenarian Editor*, p. 49.

23. Schlesinger, *op. cit.*, pp. 528–30.

24. "The Scholar in Politics," *Scribner's Monthly*, VI (1873), 608.

25. Daniel C. Gilman, *The Launching of a University* (New York, 1906), pp. 22–23. Italics in original.

26. "Darwin on the Origin of Species," *North American Review*, XC (1860), 474–506; cf. also the skeptical article, "The Origin of Species," *ibid.*, XCI (1860), 528–38.

27. Chauncey Wright, "A Physical Theory of the Universe," *North American Review*, XCIX (1864), 5.

28. "Philosophical Biology," *North American Review*, CVII (1868), 379.

29. Charles Loring Brace, "Darwinism in Germany," *North American Review*, CX (1870), 290; Chauncey Wright, "The Genesis of Species," *ibid.*, CXIII (1871), 63–103; Francis Darwin, *op. cit.*, II, 325–26. On the significance of Wright in the evolution controversy, see Sidney Ratner, "Evolution and the Rise of the Scientific Spirit in America," *Philosophy of Science*, III (1936), 104–22.

30. John Fiske, *Youmans*, p. 260.

31. See Gray, *Darwiniana, passim*.

32. *Atlantic Monthly*, XXX (1872), 507–8.

33. *Nation*, XII (1871), 258.

34. New York *Tribune*, September 19, 21, 25, 1876; cf. *Popular Science Monthly*, X (1876), 236–40.

35. See his article, "Darwinism," reprinted from the *Tribune* in *Appleton Journal*, V (1871), 350–52.

36. *Popular Science Monthly*, IV (1874), 636.

37. "Darwinism in Literature," *Galaxy*, XV (1873), 695.

38. See "Is the Religious Want of the Age Met?" *Atlantic Monthly*, XV (1860), 358–64.

39. For a characteristic exposition of orthodox views, see John T. Duffield, "Evolutionism Respecting Man, and the Bible," *Princeton Review*, LIV (1878), 150–77.

40. See Bert J. Loewenberg, "The Controversy over Evolution in New England, 1859–1873," *New England Quarterly*, VIII (1935), 232–57.

41. See John Trowbridge, "Science from the Pulpit," *Popular Science Monthly*, VI (1875), 735-36.
42. "The Darwinian Theory of the Origin of Species," *New Englander*, XXVI (1867), 607.
43. Hodge, *What Is Darwinism?* p. 7.
44. See the criticism in Asa Gray's *Darwiniana*. p. 257.
45. Hodge, *op. cit.*, pp. 52 ff., 64, 71, 177.
46. Brownson, *Works* (Detroit, 1884), IX, 265, 491-93; for Brownson's writings on the conflict between religion and science, see *ibid.*, IX, 254-331, 365-565.
47. *Christianity and Positivism* (New York, 1871), pp. 42, 63-64.
48. Fiske, *op. cit.*, p. 266.
49. *Independent*, February 23; April 12; July 16, 1868.
50. "Scientific Teaching in the Colleges," *Popular Science Monthly*, XVI (1880), 558-59.
51. Bert J. Loewenberg, in "Darwinism Comes to America 1859-1900," *Mississippi Valley Historical Review*, XXVIII (1941), 339-68, looks upon the period from 1859 to 1880 as one of probation for Darwinism, and the period from 1880 to 1900 as one that witnessed the conversion of vocal American sentiment.
52. Rev. J. M. Whiton, "Darwin and Darwinism," *New Englander*, XLII (1883), 63.
53. Gray, *Darwiniana*, pp. 176, 257, 269-70, *passim*.
54. *Religion and Science* (New York, 1873), pp. 12, 25-26.
55. Henry Ward Beecher, *Evolution and Religion* (New York, 1885), p. 51.
56. *Ibid.*, p. 52.
57. *Ibid.*, p. 115; see Paxton Hibben, *Henry Ward Beecher: An American Portrait* (New York, 1927), p. 340; E. L. Youmans, ed., *Herbert Spencer on the Americans and the Americans on Herbert Spencer*, p. 66.
58. Lyman Abbott, *The Theology of an Evolutionist* (Boston, 1897), pp. 31 ff.; see Beecher, *op. cit.*, pp. 90 ff.
59. "Agnosticism at Harvard," *Popular Science Monthly*, XIX (1881), 266; see also William M. Sloane, *The Life of James McCosh* (New York, 1896), p. 231.
60. Daniel Dorchester, *Christianity in the United States* (New York, 1888), p. 650.
61. A. V. G. Allen, *Phillips Brooks, 1835-1893* (New York, 1907), p. 309.
62. Beecher, *op. cit.*, p. 18.

CHAPTER 2

THE VOGUE OF SPENCER

1. David Duncan, *The Life and Letters of Herbert Spencer* (London, 1908), p. 128.
2. Spencer, wrote William James, "is the philosopher whom those who have no other philosopher can appreciate." *Memories and Studies*, p. 126.
3. M. De Wolfe Howe, ed., *Holmes-Pollock Letters* (Cambridge, 1941), I, 57-58. "Spencer," wrote Parrington, "laid out the broad highway over which American thought traveled in the later years of the century." *Main Currents in American Thought*, III, 198.
4. Duncan, *op. cit.*, pp. 100-1; Fisk, *The Letters of John Fiske, passim*.

5. Fiske, *Edward Livingston Youmans*, pp. 199–200. Later sales were remunerative even in England. New York *Tribune*, December 9, 1903.
6. *Atlantic Monthly*, XIV (1864), 775–76.
7. "He has so thoroughly imposed his idea," wrote John Dewey, "that even non-Spencerians must talk in his terms and adjust their problems to his statements." *Characters and Events*, I, 59–60.
8. Charles H. Cooley, "Reflections upon the Sociology of Herbert Spencer," *American Journal of Sociology*, XXVI (1920), 129. The American sociologists, Lester Ward believed as late as 1898, are "virtually disciples of Spencer." *Outlines of Sociology*, p. 192.
9. *Garrulities of an Octogenarian Editor*, p. 298. See also New York *Tribune*, December 9, 1903.
10. *Myself* (New York, 1934), p. 8.
11. Herbert Spencer, *Autobiography*, II, 113 n. The figure includes only authorized editions. Before international copyright became effective, many volumes were printed without authorization.
12. *Nation*, XXXVIII (1884), 323; see also "Another Spencer Crusher," *Popular Science Monthly*, IV (1874), 621–24. A bibliography of critical writings on Spencer is contained in J. Rumney, *Herbert Spencer's Sociology*, pp. 325–51.
13. "It appears that in the treatment of every topic, however seemingly remote from philosophy, I found occasion for falling back on some ultimate principle in the natural order." Spencer, *Autobiography*, II, 5. For an acute essay on the relation between Spencer's philosophy and his social bias see John Dewey, "Herbert Spencer," in *Characters and Events*, I, 45–62.
14. In the words of the original definition, "Evolution is an integration of matter and concomitant dissipation of motion; during which the matter passes from an indefinite, incoherent homogeneity to a definite, coherent heterogeneity; and during which the retained motion undergoes a parallel transformation." *First Principles* (4th Amer. ed., 1900), p. 407.
15. "The Instability of the Homogeneous," *ibid.*, Part II, chap. xix.
16. *Ibid.*, pp. 340–71.
17. *Ibid.*, p. 496.
18. *Ibid.*, p. 530.
19. *Ibid.*, pp. 99, 103–4.
20. Emma Brace. ed., *The Life of Charles Loring Brace*, p. 417; cf. Lyman Abbott, *The Theology of an Evolutionist*, pp. 29–30.
21. Daniel Dorchester, *Christianity in the United States*, p. 660.
22. Quoted from *The Joyful Wisdom*, in Crane Brinton, *Nietzsche* (Cambridge, 1941), p. 147.
23. *Life and Letters*, I, 68. See also *The Origin of Species*, chap. iii.
24. *My Life* (New York, 1905), pp. 232, 361.
25. "A Theory of Population, Deduced from the General Law of Animal Fertility," *Westminster Review*, LVII (1852), 468–501, esp. 499–500; "The Development Hypothesis," reprinted in *Essays* (New York, 1907), I, 1–7; see *Autobiography*, 450–51.
26 See the controversy with Weismann in Duncan, *op. cit.*, pp. 342–52.
27. *Social Statics*, pp. 79–80.
28. *Ibid.*, pp. 414-15.
29. *Ibid.*, pp. 325–444.

30. In an article on "The Relations of Biology, Psychology, and Sociology," *Popular Science*, L (1896), 163–71, Spencer defended himself against the then-common charge that his sociology had been too dependent upon biology, and argued that he had always made ample use of psychology too. In a defense of his ethical writings he also argued that he had not apotheosized the struggle for existence. "Evolutionary Ethics," *ibid.*, LII (1898), 497–502.

31. Duncan, *op. cit.*, p. 366.

32. "A Society Is an Organism," *The Principles of Sociology* (3rd ed., New York, 1925), Part II, chap. ii. For an excellent critique of Spencer's organismic theory see J. Rumney, *op. cit.*, chap. ii.

33. Spencer was not consistent in carrying out his theory of the social organism. As Ernest Barker has pointed out, he was unable to overcome the antagonism between his individualistic ethics and his organic conception of society. Barker, *Political Thought in England*, pp. 85–132. From his individualistic bias Spencer seems to have derived the atomistic idea, most clearly expressed in *Social Statics* and *The Study of Sociology*, that a society is but the sum of its individual members and takes its character from the aggregate of their characters (*Social Statics*, pp. 28–29; *The Study of Sociology*, pp. 48–51). In *The Principles of Sociology*, however, Spencer says that there arises in the social organism "a life of the whole quite unlike the lives of the units, though it is a life produced by them" (3rd ed., I, 457). A similar dualism can be found in his ethical criteria, which are sometimes determined by the impersonal requirements of evolution and sometimes by personal hedonism. Cf. A. K. Rogers, *English and American Philosophy Since 1800*, pp. 154–57.

34. *The Principles of Sociology*, II, 240–41.

35. *Ibid.*, Part V, chap. xvii.

36. The idea of the transition from militant to industrial society had greater plausibility in the period of the Pax Britannica. *Ibid.*, II, 620–28.

37. *Ibid.*, Part V, chap. xviii, "The Industrial Type of Society"; cf. *The Principles of Ethics*, Vol. II, chap. xii.

38. *Principles of Sociology*, II, 605–10; see also *Principles of Ethics*, I, 189.

39. Cooley, *op. cit.*, pp. 129–45.

40. *The Study of Sociology*, chap. i.

41. *Ibid.*, pp. 70–71.

42. Duncan, *op. cit.*, p. 367.

43. Spencer, *op. cit.*, pp. 401–2.

44. *Ibid.*, pp. 343–46.

45. "It would be strange," wrote a sociologist in 1896, "if the 'captain of the industry' did not sometimes manifest a militant spirit, for he has risen from the ranks largely because he was a better fighter than most of us. Competitive commercial life is not a flowery bed of ease, but a battle field where the 'struggle for existence' is defining the industrially 'fittest to survive.' In this country the great prizes are not found in Congress, in literature, in law, in medicine, but in industry. The successful man is praised and honored for his success. The social rewards of business prosperity, in power, in praise, and luxury, are so great as to entice men of the greatest intellectual faculties. Men of splendid abilities find in the career of a manufacturer or merchant an opportunity for the most intense energy. The very perils of the situation have a fascination

for adventurous and inventive spirits. In this fierce, though voiceless contest, a peculiar type of manhood is developed, characterized by vitality,
energy, concentration, skill in combining numerous forces for an end, and
great foresight into the consequence of social events." C. R. Henderson,
"Business Men and Social Theorists," *American Journal of Sociology,*
I (1896), 385–86.
46. *My Memories of Eighty Years* (New York, 1922), pp. 383–84.
47. *Highways of Progress* (New York, 1910), p. 126; cf. also p. 137.
48. Quoted in William J. Ghent, *Our Benevolent Feudalism,* p. 29.
49. *Autobiography of Andrew Carnegie* (Boston, 1920), p. 327.
50. "Wealth," *North American Review,* CXLVIII (1889), 655–57.
51. Allan Nevins, ed., *Selected Writings of Abram S. Hewitt* (New York,
1910), p. 277.
52. Lochner *v.* New York, 198 U.S. 45 (1905).
53. Youmans, "The Recent Strike," *Popular Science Monthly,* III (1872),
623–24. See also R. G. Eccles, "The Labor Question," *ibid.,* XI (1877),
606–11; *Appleton's Journal,* N. S., V (1878), 473–75.
54. "The Social Science Association," *Popular Science Monthly,* V (1874),
267–69. See also *ibid.,* VII (1875), 365–67.
55. "On the Scientific Study of Human Nature," reprinted in Fiske, *op. cit.,*
p. 482. For other statements of the conservative Spencerian viewpoint,
see Erastus B. Bigelow, "The Relations of Capital and Labor," *Atlantic
Monthly,* XLII (1878), 475–87; G. F. Parsons, "The Labor Question,"
ibid., LVIII (1886), 97–113. Also "Editor's Table," *Appleton's Journal,*
N. S., V (1878), 473–75.
56. Henry George, *A Perplexed Philosopher,* pp. 163–64 n. Fiske shared Youmans' conservatism, but was less alarmed at the menace of radicalism to
the American future. See Fiske, *op. cit.,* pp. 381–82 n. For the social
outlook of an American thinker thoroughly influenced by Spencer, see
Henry Holt, *The Civic Relations* (Boston, 1907), and *Garrulities of an
Octogenarian Editor,* pp. 374–88.
57. Duncan, *op. cit.,* p. 225.
58. Youmans, ed., *Herbert Spencer on the Americans,* pp. 9–20.
59. *Ibid.,* pp. 19–20. See *Nation,* XXXV (1882), 348–49.
60. Spencer, *Autobiography,* p. 479.
61. Burton J. Hendrick, *The Life of Andrew Carnegie* (New York, 1932),
I, 240.
62. *Ibid.,* Vol. II, chap. xii.
63. W. H. Hudson, "Herbert Spencer," *North American Review,* CLXXVIII
(1904), 1–9.
64. *Catholic World,* cited in *Current Opinion,* LXIII (1917), 263.
65. These essays were republished in book form in 1916 as *The Man Versus
the State,* edited by Truxton Beale.
66. *Ibid.,* p. ix. See comment in *Nation,* CI (1915), 538. Albert Jay Nock
published in 1940, as a critique of the New Deal, a volume of Spencer's
essays under the same title as Beale's edition.
67. See Thomas C. Cochran, "The Faith of Our Fathers," *Frontiers of Democracy,* VI (1939), 17–19.
68. Robert S. and Helen M. Lynd, *Middletown in Transition* (New York,
1937), p. 500.

CHAPTER 3

WILLIAM GRAHAM SUMNER: SOCIAL DARWINIST

1. Charles Page, *Class and American Sociology*, pp. 74, 103, has stressed the importance of the economic ethics of the Protestant tradition as an element in Sumner's thinking. Ralph H. Gabriel has pointed out its importance in American thought during Sumner's period in *The Course of American Democratic Thought*, pp. 147–60. Passages in Sumner's writings illustrative of this tradition may be found in *Essays of William Graham Sumner*, edited by A. G. Keller and M. R. Davie, II, 22 ff., and *The Challenge of Facts and Other Essays*, pp. 52, 67.

2. *Essays*, II, 22.

3. *Earth-Hunger and Other Essays*, p. 3.

4. *Illustrations of Political Economy* (London, 1834), III, Part I, 134–35, and Part II, 130–31; VI, Part I, 140, and Part II, 143–44.

5. *The Challenge of Facts*, p. 5.

6. Harris E. Starr, *William Graham Sumner*, pp. 47–48.

7. Cf. Albert Galloway Keller's discussion of Sumner's influence in "The Discoverer of the Forgotten Man," *American Mercury*, XXVII (1932), 257–70.

8. William Lyon Phelps, "When Yale Was Given to Sumnerology," *Literary Digest International Book Review*, III (1925), 661–63.

9. *Ibid.*, p. 661.

10. Starr, *op. cit.*, p. 322.

11. Cf. *What Social Classes Owe to Each Other*, pp. 155–56.

12. Cf. the preface to *The Science of Society*, I, xxxiii. Sumner died before the completion of this work, and it was finished by Albert Galloway Keller and published in four volumes in 1927 by the Yale University Press.

13. See the autobiographical sketch in *The Challenge of Facts*, p. 9. Keller, estimating the major influences on Sumner's sociology, has placed Spencer first, Julius Lippert second, and Gustav Ratzenhofer third. "William Graham Sumner," *American Journal of Sociology*, XV (1910), 832–35. Lippert was a German cultural historian whose method was much like that employed in *Folkways*. See his *Kulturgeschichte der Menschenheit* (1886), translated in 1931 by George Murdock as *The Evolution of Culture*. Ratzenhofer was a German sociologist of the conflict school.

Sumner, of course, was not an uncritical follower of Spencer. Sumner did not accept Spencer's identification of evolution with progress, and Spencer's optimism was meaningless to him. He was not so severe in his conception of the proper limitations of government. Cf. Starr, *op. cit.*, pp. 292–93. Less libertarian, he understood the limitations imposed by industrial society upon individual freedom. *Essays*, I, 310 ff. Finally, his ethical relativism was unlike Spencer's ethical theory.

For his part, Spencer cordially approved Sumner's way of defending laissez faire and property rights. He tried to persuade the Liberty and Property Defense League in England to reprint *What Social Classes Owe to Each Other*. Starr, *op. cit.*, pp. 503–5.

14. *Science of Society*, chap. i; cf. also the essay "Earth-Hunger." The main elements of this idea are similar to the wage-fund doctrine and can be traced to Sumner's early acquaintance with Harriet Martineau.

15. *What Social Classes Owe to Each Other,* p. 17; cf. also p. 70. "Nature is entirely neutral; she submits to him who most energetically and resolutely assails her. She grants her rewards to the fittest . . . without regard to other considerations of any kind. If there be liberty, men get from her just in proportion to their being and their doing." *The Challenge of Facts,* p. 25.

16. *What Social Classes Owe to Each Other,* p. 76.

17. At times Sumner distinguished the struggle for existence, which he looked upon as man's impersonal struggle against nature, from what he called "the competition of life," a strictly social form of conflict, in which groups of men united in the conquest-of-nature struggle among themselves. Cf. *Folkways,* pp. 16–17, and *Essays,* I, 142 ff.

18. *Essays,* II, 56.

19. *The Challenge of Facts,* p. 68.

20. *Ibid.,* pp. 40, 145–50; *Essays,* I, 231.

21. *The Challenge of Facts,* pp. 43–44.

22. *What Social Classes Owe to Each Other,* p. 73.

23. *Essays,* I, 289.

24. *What Social Classes Owe to Each Other,* pp. 54–56.

25. *The Challenge of Facts,* p. 90.

26. *The Science of Society,* I, 615. Cf. also p. 328, where Sumner opposes a communal economy on the ground that it makes variation impossible — "and variation is the starting-point of new adjustment." Sumner considered the masses to be immobile and unproductive of social improvement. Variation is chiefly characteristic of the upper classes. *Folkways,* pp. 45–47.

27. *The Challenge of Facts,* p. 67.

28. *What Social Classes Owe to Each Other,* p. 135.

29. *Folkways,* p. 48.

30. *Essays,* I, 358–62.

31. *Ibid.,* I, 86–87.

32. *Earth Hunger,* pp. 283–317.

33. *Essays,* I, 185.

34. *Ibid.,* I, 104.

35. *Ibid.,* II, 165.

36. See "Advancing Organization in America," *ibid.,* II, 340 ff., especially 349–50. In his references to the effects of the frontier upon the unique historical development of the United States, Sumner seems to have anticipated also the theories of Frederick Jackson Turner. Sumner's views on democracy have been discussed in Gabriel, *op. cit.,* chap. xix, and by Harry Elmer Barnes in "Two Representative Contributions of Sociology to Political Theory: The Doctrines of William Graham Sumner and Lester Frank Ward," *American Journal of Sociology,* XXV (1919), 1–23, 150–70.

37. "The Absurd Effort to Make the World Over," in *Essays,* I, 105.

38. *Ibid.,* II, 215.

39. See "Reply to a Socialist," in *The Challenge of Facts,* pp. 58, 219; on the ineffectiveness of reform legislation, see *War and Other Essays,* pp. 208–310; *Earth-Hunger,* pp. 283 ff.; and *What Social Classes Owe to Each Other,* pp. 160–61.

40. *The Challenge of Facts,* p. 57.

41. *Essays*, I, 109.
42. *What Social Classes Owe to Each Other*, p. 101.
43. *Essays*, II, 249-53, 255.
44. *Ibid.*, II, 67-76.
45. *The Challenge of Facts*, pp. 27-28.
46. *Ibid.*, p. 99; *What Social Classes Owe to Each Other*, pp. 90-95.
47. *Essays*, II, 366.
48. *Ibid.*, II, 435.
49. Starr, *op. cit.*, pp. 285-88; cf. *What Social Classes Owe to Each Other*, p. 146.
50. *The Goose-Step* (Pasadena, 1924), p. 123.
51. Starr, *op. cit.*, pp. 258, 297.
52. See the essays on democracy and plutocracy in *Essays*, II, 213 ff.
53. *Ibid.*, II, 236-37.
54. " The Forgotten Man," *ibid.*, I, 466-96; cf. also *What Social Classes Owe to Each Other, passim.*
55. *The Challenge of Facts*, p. 74.
56. Starr, *op. cit.*, p. 275.
57. Phelps, *op. cit.*, p. 662.
58. Quoted in Starr, *op. cit.*, pp. 300-1.
59. For evidence that this aspect of Sumner's thought is by no means dead, however, see some of the comments in *Sumner Today* (New Haven, 1940), ed. Maurice R. Davie.
60. *Folkways*, pp. 4, 29.
61. Cf. the review of *Folkways* by George Vincent in *American Journal of Sociology*, XIII (1907), 414-19; also John Chamberlain, " Sumner's Folkways," *New Republic*, IC (1939), 95.

CHAPTER 4

LESTER WARD: CRITIC

1. See the discussion of Comte in Ludwig Gumplowicz, *The Outlines of Sociology*, pp. 28-29.
2. Cited in Edward A. Ross, *Foundations of Sociology*, p. 48.
3. *Dynamic Sociology*, I, 6; cf. pp. 142-44. See also *The Psychic Factors of Civilization*, p. 2.
4. *Glimpses of the Cosmos*, I, xx-xxi; VI, 143.
5. Biographical material on Ward may be found in Emily Palmer Cape, *Lester F. Ward*; Bernhard J. Stern, *Young Ward's Diary*; and scattered throughout the six volumes of *Glimpses of the Cosmos*.
6. See George A. Lundberg, *et al., Trends in American Sociology*, chap. i.
7. Howard W. Odum, ed., *American Masters of Social Science* (New York, 1927), p. 95.
8. On the neglect of Ward, see Samuel Chugerman, *Lester Ward, The American Aristotle* (Durham, 1939), chap. iii.
9. Richard T. Ely to Ward, November 22, 1887, Ward MSS, Autograph Letters, II, 35; Ely to Ward, July 30, 1890, *ibid.*, III, 48; " The Letters of Albion W. Small to Lester F. Ward," Bernhard J. Stern, ed., *Social Forces*, XII (1933), 164-65.
10. See " Broadening the Way to Success," *Forum*, II (1886), 340-50; " The

Use and Abuse of Wealth," *ibid.*, III (1887), 364–72; "Plutocracy and Paternalism," *ibid.*, XX (1895), 300–10. The Ward MSS have much unique material which sheds light on the range of Ward's influence.

11. See the autobiographical remarks in *Applied Sociology*, pp. 105–6, 127–28.
12. *Glimpses*, II, 164–71.
13. *Ibid.*, II, 336–37.
14. *Ibid.*, II, 342–45.
15. *Ibid.*, p. 352.
16. *Glimpses*, III, 45–47; see also VI, 58–63.
17. *Ibid.*, IV, 350–63; cf. *The Psychic Factors of Civilization*, chap. xxxiii; *Pure Sociology*, p. 16. Views similar to Ward's on the limits of the value of competition in human affairs and the uniquely rational character of human evolution were advocated by his friend Major John W. Powell, first chief of the U.S. Bureau of Ethnology. See Powell's "Competition as a Factor in Human Evolution." *The American Anthropologist*, I (1888), 297–323; and "Three Methods of Evolution," *Bulletin*, Philosophical Society of Washington, VI (1884), xlvii–lii.
18. *Dynamic Sociology*, I, v–vi.
19. *Ibid.*, I, 468.
20. *Ibid.*, I, 15–16, 29–30.
21. *Ibid.*, I, 706.
22. *Ibid.*, II, chaps. ix–xii.
23. *Dynamic Sociology*, II, 539.
24. Ward's views on education are discussed fully by Elsa P. Kimball in *Sociology and Education*.
25. See *Glimpses*, III, 147–48.
26. *Ibid.*, IV, 246–52; see also Ward's discussion of Lamarck and neo-Darwinism, *ibid.*, IV, 253–95.
27. *Pure Sociology*, p. 204.
28. *Ibid.*, p. 204.
29. *Ibid.*, pp. 237–40.
30. *Ibid.*, pp. 215–16.
31. Bernhard J. Stern, ed., "The Letters of Ludwig Gumplowicz to Lester F. Ward," *Sociologus*, I (1933), 3–4.
32. *The Psychic Factors of Civilization*, chap. xi; *Outlines of Sociology*, p. 27.
33. *The Psychic Factors of Civilization*, pp. 134–35.
34. *Glimpses*, III, 303–4. See also Ward's review of Giddings' *Principles of Sociology*, *ibid.*, V, 282–305.
35. *Ibid.*, V, 38–66.
36. *Outlines of Sociology*, p. 61; see "Herbert Spencer's Sociology" in *Glimpses*, VI, 169–77.
37. *The Psychic Factors of Civilization*, pp. 298–99.
38. *Ibid.*, p. 100.
39. See "Politico-Social Functions," in *Glimpses*, II, 336–46.
40. *Dynamic Sociology*, II, 576–83.
41. *Ibid.*, I, 104, 137, 50.
42. Stern, ed., *op. cit.*, XV (1937), 318, 320; "The Ward-Ross Correspondence, 1891–1896,' *American Sociological Review*, III (1938), 399.
43. "Social Darwinism," *American Journal of Sociology*, XII (1907), 710.
44. *Applied Sociology*, chap. xiii; *Outlines of Sociology*, pp. 273 ff., 292–93; *The Psychic Factors of Civilization*, chap. xxxviii.

45. *Outlines of Sociology,* p. 293.
46. Stern, ed., *op. cit.,* XV, 313.

CHAPTER 5

EVOLUTION, ETHICS, AND SOCIETY

1. See Robert H. Lowie, *The History of Ethnological Theory,* pp. 20 ff.
2. *Democracy* (New York, 1925), p. 78.
3. *The Principles of Sociology,* II, 240–41. An interesting offshoot of the conflict emphasis was John Stahl Patterson's anonymously published *Conflict in Nature and Life* (New York, 1883).
4. Emma Brace, *Charles Loring Brace,* p. 365.
5. "What Morality Have We Left?" *North American Review,* CXXXII (1881), 504.
6. Quoted in Joseph Dorfman, *Thorstein Veblen and His America,* p. 46
7. "The Prospect of a Moral Interregnum," *Atlantic Monthly,* XLIII (1879), 629–42, esp. 636.
8. "Malthusianism, Darwinism, and Pessimism," *North American Review,* CXXIX (1879), 447–72. Bowen did not doubt the social-Darwinian premise that the upper classes are in some way equal to the fit.
9. M. A. Hardaker, "A Study in Sociology," *Atlantic Monthly,* L (1882), 214–20.
10. *A Traveler from Altruria* (New York, 1894), pp. 12–13.
11. Titus M. Coan, "Zealot and Student," *Galaxy,* XX (1875), 177, 183; G. C. Eggleston, "Is the World Overcrowded?" *Appeton's Journal,* XIV (1875), 530–33; N. S. Shaler, "The Uses of Numbers in Society," *Atlantic Monthly,* XLIV (1879), 321–33.
12. *The Descent of Man* (London, 1874), pp. 151–52.
13. *Ibid.,* pp. 706–7. See Geoffrey West, *Charles Darwin* (New Haven, 1938), pp. 327–28.
14. *The Descent of Man,* chaps. iv, v. A good study of Darwin's views on the role of mutual aid and moral law in human progress may be found in George Nasmyth, *Social Progress and the Darwinian Theory,* chap. ix.
15. *Mutual Aid* (New York, 1902), chap. i.
16. W. Bagehot, *Physics and Politics, passim,* esp. pp. 24, 36–37, 40–43, 64.
17. *The Meaning of Infancy* (1883); *Outlines of Cosmic Philosophy* (13th ed., 1892), II, 342 ff. That the winds of doctrine were blowing steadily in the direction of the evolutionists may be seen by comparing the great popularity of Fiske with the neglect of Jacob Gould Schurman's little book, *The Ethical Import of Darwinism.* Schurman, who was Sage Professor of Philosophy at Cornell University, tried to show that Darwinism did not logically undermine the traditional sanctions of conduct because they were rooted in something more than natural evolution. Attempting to put Darwin in his historical setting, Schurman pointed out that his theory had a logical affinity with the doctrines of the Utilitarians as well as Malthus. All of Darwin's natural selection, Schurman argued, on the basis of some interesting textual evidence, is based upon the preconceptions of utilitarianism — the survival of *useful,* or as Darwin called them, "profitable" variations in the organism. Schurman objected to the tendency of evolutionists in ethics to assume that, because natural selection presupposes a utility, morality is nothing but a utility. He concluded

that morals could be firmly founded only on an intuitionist basis. *The Ethical Import of Darwinism*, pp. 116 ff., 141–60, *passim*. An idealistic attack on Spencer's ethics was made by James Thompson Bixby in *The Ethics of Evolution* (New York, 1891). A comprehensive review of the literature was made by C. M. Williams, *A Review of the Systems of Ethics Founded on the Theory of Evolution* (New York, 1893).

18. *Evolution and Ethics and Other Essays* (1920), pp. 36–37.
19. *Evolution and Ethics and Other Essays*, pp. 44–45. The main essay is on pp. 46–116; the argument of the essay is expanded in a " Prolegomena," on pp. 1–45.
20. *The Ascent of Man*, p. 36.
21. *Ibid.*, p. 211.
22. *Mutual Aid*, pp. 74–75.
23. See Benjamin Kidd, *Social Evolution*, pp. 72–73.
24. *Ibid.*, pp. 36–37.
25. *Ibid.*, p. 68.
26. *Ibid.*, chap. iv.
27. *Ibid.*, chap. viii.
28. Albion W. Small, in Stern, ed., *op. cit.*, XII, 170.
29. " Mr. Kidd's Social Evolution," *American Journal of Sociology*, I (1895), 311–12.
30. " Kidd's Social Evolution," *North American Review*, CLXI (1895), 94–109.
31. *Aristocracy and Evolution* (London, 1898), *passim*.
32. Drummond and Kropotkin were aware of the extent of their agreement. Drummond acknowledged his indebtedness to Fiske and Kropotkin (*Ascent of Man*, pp. 239–40, 282–83), while Kropotkin returned the compliment and also mentioned Giddings' principle of " the consciousness of kind." *Mutual Aid*, p. xviii.

CHAPTER 6

THE DISSENTERS

1. The emphasis in this chapter on urban movements and thinkers arises from no desire to minimize the importance of agrarian protest to American radicalism. Organized grass-roots movements, however, showed little interest in anything resembling a systematic theory of society.
2. On the history and ideology of the social-gospel movement, the author is deeply indebted to Charles Howard Hopkins' *The Rise of the Social Gospel in American Protestantism, 1865–1915*. See also James Dombrowski, *The Early Days of Christian Socialism in America*, which contains (in the first chapter) an analysis of the ideology of the social gospel. For an informative contemporary discussion, see Nicholas Paine Gilman, *Socialism and the American Spirit* (London, 1893).
3. Rauschenbusch, *Christianizing the Social Order*, p. 9.
4. *Ibid.*, p. 90.
5. See George Herron, *Between Caesar and Jesus* (New York, 1899), pp. 45 ff.
6. *Christianity and Social Problems* (Boston, 1896), p. 133.
7. Behrends, *Socialism and Christianity*, p. 6. See also Lyman Abbott, *op. cit.*, p. 120; Gladden, *Social Facts and Forces* (New York, 1897), p. 2;

Tools and the Man (Boston, 1893), p. 3; Josiah Strong, *The Next Great Awakening* (New York, 1902), pp. 171–72.

8. Behrends, *op. cit.*, pp. 64–66.

9. *Applied Christianity*, pp. 104–5; cf. pp. 111–12, 130. The whole object of the Christian scheme of ethics, Gladden believed, is to counteract injuries wrought by the survival of the fittest. Gladden, *Tools and the Man*, pp. 275–78.

10. *Ibid.*, p. 36.

11. *Ibid.*, p. 176; cf. pp. 270, 287–88. See also *Ruling Ideas of the Present Age* (Boston, 1895), pp. 63 ff., 73–74, 107; *Social Facts and Forces*, pp. 93, 220; *Recollections* (Boston, 1909), p. 419.

12. *The Christian Society* (New York, 1894), pp. 103, 108–9.

13. *The Christian State* (New York, 1895), p. 88; *The New Redemption* (New York, 1893), pp. 16–17. For the attitude of Rauschenbusch toward competition, see *Christianity and the Social Crisis*, pp. 308 ff., and *Christianizing the Social Order, passim*.

14. *The New Redemption*, p. 30.

15. See Gladden, *Ruling Ideas of the Present Age*, p. 107; *Tools and the Man*, p. 176; Herron, *The Christian State*, p. 88; Josiah Strong, *op. cit.*, pp. 171–72.

16. See *The Science of Political Economy* (New York, 1897), pp. 402–3.

17. *Progress and Poverty* (New York, 1879), p. 104.

18. *Progress and Poverty*, pp. 342–43.

19. *Ibid.*, pp. 344–49.

20. *Ibid.*, p. 349–90.

21. *A Perplexed Philosopher*, p. 87. See Henry George, Jr., *The Life of Henry George*, pp. 369–70, 420, 568 ff.

22. *Looking Backward* (1889), pp. 60–61.

23. *Ibid.*, p. 244. A more detailed analysis of nineteenth-century capitalism was presented by Bellamy in his *Equality*.

24. *Nationalist*, I (1889), inside cover page.

25. *Edward Bellamy Speaks Again!* pp. 34–35.

26. See *Nationalist*, I (1889), 55–57; II (1890), 61–63, 135–38, 155–62.

27. Ward, *Glimpses*, IV, 346. See the letter from M. A. Clancy, secretary of the Nationalist Club of Washington, to Ward, February 23, 1889, Ward MSS, Autograph Letters, III, 18.

28. *The Coöperative Commonwealth*, pp. 40, 77–83, 88.

29. *Our Destiny*, pp. 13–14, 18–22, 36–37, 73, 86–95, 113–14; cf. *The Coöperative Commonwealth*, pp. 171–72, 179, 220. In a subsequent volume, Gronlund reviewed the mistakes of the Bryan campaign and renewed his plea for the acceptance and collectivization of the trusts. *The New Economy, passim*.

30. *The Correspondence of Marx and Engels* (New York, 1935), pp. 125–26.

31. See the preface to Lewis' *Ten Blind Leaders of the Blind* (Chicago, 1909), p. 3.

32. Raphael Buck, "Natural Selection Under Socialism," *International Socialist Review*, II (1902), 790. See also Robert Rives La Monte, "Science and Socialism," *ibid.*, I (1900), 160–73; Herman Whitaker, "Weismannism and Its Relation to Socialism," *ibid.*, I (1901), 513–23; J. W. Sumners, "Socialism and Science," *ibid.*, II (1902), 740–48; A. M. Simons, "Kropotkin's 'Mutual Aid,'" *ibid.*, III (1903), 344–49.

33. Robert Rives La Monte, *Socialism, Positive and Negative* (Chicago, 1902), pp. 18–19; A. M. Lewis, *An Introduction to Sociology*, pp. 173–87.

34. See A. M. Lewis, *Evolution, Social and Organic*, chaps. vii and ix. Socialist intellectuals in this country and in Europe drew heavily upon Enrico Ferri's *Socialism and Modern Science*. See also Ernest Untermann, *Science and Revolution* (Chicago, 1905), chap. xv. Cf. A. M. Lewis, *op. cit.,* chap. vii, " A Reply to Haeckel." See also Anton Pannekoek, *Marxism and Darwinism* (Chicago, 1912).

35. Lewis, *op. cit.,* pp. 60–80, esp. p. 78. See also, Herman Whittaker, *op. cit.*

36. Lewis, *op. cit.,* pp. 81–96, esp. pp. 93–95; W. J. Ghent, *Socialism and Success* (New York, 1910), pp. 47–49.

37. Lewis, *op. cit.,* pp. 97–114, 168–82. Cf. *An Introduction to Sociology, passim.*

38. *The Larger Aspects of Socialism,* p. 86. For the whole of Walling's arguments on these points see chaps. i–iv.

39. The most acute early diagnosis of this change was W. J. Ghent's *Our Benevolent Feudalism.*

40. *Wealth Against Commonwealth,* pp. 494–95.

41. See above, Chapter 4, note 17.

42. *The New Freedom* (New York, 1914), p. 15.

43. *The New Democracy,* pp. 49–50.

44. " Shall We Abandon the Policy of Competition? " reprinted in *The Curse of Bigness,* p. 104.

45. Humanitarian discontent with the scientific apology for the elimination of the unfit was expressed by Charlotte Perkins Gilman, a prominent social worker:

> " Then science comes with solemn air, and shows us social laws,
> Explaining how the poor are there from a purely natural cause.
> 'Tis natural for low and high to struggle and to strive;
> 'Tis natural for the worse to die and the better to survive.

> " We swallowed all this soothing stuff, and easily were led
> To think if we were stern enough, the poor would soon be dead.
> But, O! in vain we squeeze, grind, and drive them to the wall —
> For all our deadly work we find it does not kill them all!

> " The more we struggle they survive; increase and multiply!
> There seem to be more poor alive, in spite of all that die!
> Whene'er I take my walks abroad how many poor I see,
> And eke at home! How long, O Lord! How long must this thing be! "
> *In This Our World* (Boston, 1893), pp. 201–2.

46. See the dissenting opinion, Plant *v.* Woods, 176 Mass. 492 (1900), quoted in *Representative Opinions of Mr. Justice Holmes* (New York, 1931), p. 316.

47. *Drift and Mastery,* p. 267; cf. Wilson, *op. cit.,* p. 20.

48. It is significant that Wilson invoked the social organism to justify state interference under the Constitution. " All that progressives ask or desire," he declared, " is permission — in an era when ' development,' ' evolution,' is the scientific word — to interpret the Constitution according to the Darwinian principle; all they ask is recognition of the fact that a nation

is a living thing and not a machine." Wilson, *op. cit.*, pp. 44–48. Wilson's volume, *The State* (Boston, 1889), was considerably colored by Darwinian concepts.

49. For a graphic presentation of the magnitude of the state apparatus of the 1930's, see Louis M. Hacker, *American Problems of Today* (New York, 1938), pp. 276–81.

50. Quoted in Ralph H. Gabriel, *The Course of American Democratic Thought*, p. 306, from Henry A. Wallace, *Statesmanship and Religion* (1934).

CHAPTER 7
THE CURRENT OF PRAGMATISM

1. For an account of the social teachings of Harris, see Merle Curti, *The Social Ideas of American Educators*, chap. ix.

2. See John Dewey, *Experience and Nature* (Chicago, 1926), pp. 282–83.

3. "There can be little doubt," writes Morris R. Cohen, "that Peirce was led to the formulation of the principle of pragmatism through the influence of Chauncey Wright." Charles Peirce, *Chance, Love, and Logic*, pp. xviii–xix. On Wright see Gail Kennedy, "The Pragmatic Naturalism of Chauncey Wright," *Columbia University Studies in the History of Ideas*, III (1935), 477–503; Ralph Barton Perry, *The Thought and Character of William James*, I, chap. xxxi; William James, "Chauncey Wright," in *Collected Essays and Reviews*, pp. 20–25; Sidney Ratner, "Evolution and the Rise of the Scientific Spirit in America," *Philosophy of Science*, III, 104–22. Wright's most significant essays have been collected by Charles Eliot Norton in *Philosophical Discussions*.

4. *Philosophical Discussions*, p. 56.

5. The clearest interpretation of Wright's view is that of Morris R. Cohen in *The Cambridge History of American Literature* (New York, 1917–23), III, 236.

6. Peirce, *op. cit.*, p. 190.

7. *Ibid.*, p. 162.

8. *Ibid.*, pp. 162–63; *Collected Papers* (Cambridge, 1931–35), VI, 51–52.

9. *Chance, Love, and Logic*, p. 45. Peirce in his essay did not expound the pragmatic test as one of the truth of ideas but only as one of their clarity. For the differences between Peirce's and James's pragmatism see the essay by John Dewey, *ibid.*, pp. 301–8, and Justus Buchler, *Charles Peirce's Empiricism* (New York, 1939), pp. 166–74.

Peirce's objection to what he thought was the ethical implication of Darwinism is worth noting. *The Origin of Species*, he asserted in 1893, "merely extends politico-economical views of progress to the entire realm of animal and vegetable life. . . . Among animals the mere mechanical individualism is vastly reënforced as a power making for good by the animal's ruthless greed. As Darwin puts it on his title-page, it is the struggle for existence; and he should have added for his motto: Every individual for himself, and the Devil take the hindmost. Jesus, in his sermon on the Mount, expressed a different opinion." Peirce, *Chance, Love, and Logic*, p. 275.

10. See Perry, *op. cit.*, I, *passim*. C. Hartley Grattan, *The Three Jameses, A Family of Minds*.

11. Perry, *op. cit.*, I, 320–23. For the influence of the French thinker, Charles Renouvier, on James at this time, see Perry and also *The Will to Believe*, p 143; *Some Problems of Philosophy* (New York, 1911), pp. 163–65.
12. *Some Problems of Philosophy*, p. 165 n.
13. See John Dewey, " William James," in *Characters and Events*, I, 114–15; Theodore Flournoy, *The Philosophy of William James* (New York, 1917), pp. 34–35, 112, 144–45.
14. *A Pluralistic Universe* (London, 1909), pp. 49–50; cf. *Some Problems of Philosophy*, pp. 142–43.
15. *Memories and Studies*, pp. 127–28.
16. Perry, *op. cit.*, I, 482.
17. " Herbert Spencer," *Nation*, LXXVII (1903), 460.
18. *Memories and Studies*, p. 112.
19. *Pragmatism*, p. 39.
20. Perry, *op. cit.*, I, 482–83. The parody, which has been credited to James, was first rendered by the English mathematician Thomas Kirkman in his *Philosophy Without Assumptions* (London, 1876), p. 292. See the appendix to the fourth American edition of Spencer's *First Principles*, esp. pp. 577–83.
21. *Pragmatism*, pp. 105–6.
22. Perry, *op. cit.*, I, 486–87.
23. *Pragmatism*, pp. 23–33.
24. *The Will to Believe*, pp. 161–66.
25. *Collected Essays and Reviews*, pp. 148–49. Compare the statement of John Dewey in *The Influence of Darwin on Philosophy*, pp. 16–17.
26. *Collected Essays and Reviews*, p. 67.
27. See, for example, chap. xi on " Attention."
28. *Principles of Psychology*, I, 140–41.
29. *Pragmatism*, p. 201.
30. *Atlantic Monthly*, XLVI (1880), 441–59; reprinted in *The Will to Believe*, pp. 216–54. See also the companion piece, " The Importance of Individuals," *Open Court*, IV (1890), 2437–40, reprinted in *The Will to Believe*, pp. 255–62, and the answers to James by John Fiske and Grant Allen in the *Atlantic Monthly*, XLVII (1881), 75–84, 371–81.
31. *The Will to Believe*, pp. 253–54.
32. *Ibid.*, pp. 257–58, 262. Cf. John Dewey in *The Quest for Certainty*, p. 244: " If existence were either completely necessary or completely contingent, there would be neither comedy nor tragedy in life, nor need of the will to live."
33. James's individualism is stressed in Curti, *The Social Ideas of American Educators*, chap. xiii. For James's social views in general and his interest in reforms, see Perry, *op. cit.*, vol. II, chaps. lxvii, lxviii.
34. *The Will to Believe*, pp. 160–61.
35. *The Letters of William James*, I, 284.
36. *Loc. cit.*
37. *Memories and Studies*, pp. 140–41
38. " Herbert Spencer," *Nation*, LXXVII (1903), 461.
39. *The Letters of William James*, I, 252.
40. *Ibid.*, II, 318. James seems also to have been influenced by the writings of H. G. Wells.
41. " The Moral Equivalent of War," in *Memories and Studies*, p. 286.

42. *Talks to Teachers on Psychology: and to Students on Some of Life's Ideals* (New York, 1925), pp. 298–99.

43. *The Letters of William James*, II, 201.

44. See John Dewey, "From Absolutism to Experimentalism," in George P. Adams and William P. Montague, eds., *Contemporary American Philosophy* (New York, 1930), pp. 23–24; also the biographical chapter edited by Jane Dewey in Paul A. Schilpp, ed., *The Philosophy of John Dewey*, pp. 3–45.

45. The great scope and variety of Dewey's work, and also the contextual nature of his thought, make any attempt to portray his impact on the ideas recorded here necessarily fragmentary.

46. On the element of Darwinism in Dewey's approach to knowledge and its limitations in accounting for his theory of knowledge, see W. T. Feldman, *The Philosophy of John Dewey* (Baltimore, 1934), chaps. iv, vii.

47. *Reconstruction in Philosophy*, pp. 84–86; *Essays in Experimental Logic* (Chicago, 1916), pp. 331–32

48. "The Interpretation of Savage Mind," *Psychological Review*, IX (1902), 219.

49. "The Need for Social Psychology," *ibid.*, XXIV (1917), 273.

50. *The Quest for Certainty;* p. 244 and chap. ix; *Experience and Nature*, esp. pp. 62–77; *Human Nature and Conduct*, pp. 308–11.

51. For an early statement on determinism and ethics, see *The Study of Ethics* (Ann Arbor, 1894), pp. 132–38.

52. *Reconstruction in Philosophy*, *passim*, esp. pp. 125–26; *The Influence of Darwin on Philosophy and Other Essays*, pp. 17, 271–304, esp. pp. 273–74.

53. Adams and Montague, eds., *op. cit.*, p. 20. See Dewey's article on "The Ethics of Democracy," University of Michigan *Philosophical Papers*, Second Series (1888).

54. "Social Psychology," *Psychological Review*, I (1894), 400–9.

55. *The Quest for Certainty*, pp. 211–12. For Dewey's historical analysis of Spencer's individualism, see *Characters and Events*, I, 52 ff.

56. *The Public and Its Problems*, pp. 73–74.

57. *Characters and Events*, II, 728–29.

58. In 1897 Dewey stated his belief that "education is the fundamental method of social progress and reform" and that "every teacher . . . is a social servant set apart for the maintenance of proper social order and the securing of the right social growth." "My Pedagogic Creed," *Teachers' Manuals*, No. 25 (New York, 1897), pp. 16, 18. In *Democracy and Education* he conceived of education as selective environment and argued its possibilities as a means of social change; see esp. chap. ii. Cf. Curti, *op. cit.*, chap. xv; Sidney Hook, *John Dewey, An Intellectual Portrait* (New York, 1939), chap. ix.

59. Adams and Montague, eds., *op. cit.*, p. 23.

60. "Evolution and Ethics," *Monist*, VIII (1898), 321–41. Dewey's articles on "The Evolutionary Method as Applied to Morality," *Philosophical Review*, XI (1902), 109–24, 353–71, illustrate his conception of the significance of the genetic method for ethics.

61. Dewey and Tufts, *Ethics*, pp. 368–75.

62. See *Characters and Events*, I, 121–22; II, 435–42, 542–47; "The Development of American Pragmatism," *Studies in the History of Ideas*, Department of Philosophy, Columbia University, II (1925), 374.

CHAPTER 8
TRENDS IN SOCIAL THEORY, 1890–1915

1. " Recent Progress of Political Economy in the United States," *Publications,* American Economic Association, IV (1889), 26.
2. A more limited parallel than that offered here was made by John M. Keynes, *Laissez-Faire and Communism* (New York, 1926), pp. 39–43.
3. " The Preconceptions of Economic Science," Part III, *Quarterly Journal of Economics,* XIV (1900), 257 n.
4. One did not have to be an orthodox follower of the classical school to appreciate the intellectual security of this way of thinking. " Rightly viewed, perfect competition would be seen to be the order of the economic universe, as truly as gravity is the order of the physical universe, and to be not less harmonious and beneficent in operation." Francis A. Walker, *Political Economy* (3rd ed., New York, 1888), p. 263. For a good collation of the clichés of natural-law economics, see Henry Wood, *The Political Economy of Natural Law* (Boston, 1894). Cf. also John B. Clark, *Essentials of Economic Theory* (New York, 1907).
5. Francis A. Walker, *The Wages Question,* pp. 240–41 n.
6. *Ibid.,* p. 142.
7. Francis Wayland, *The Elements of Political Economy,* recast by Aaron L. Chapin (New York, ed. 1883), pp. i, 4–6, 174; Francis Bowen, *American Political Economy* (New York, ed. 1887), p. 18; Arthur Latham Perry, *Introduction to Political Economy* (New York, 1880), pp. 52, 60, 75, 100; J. Laurence Laughlin, *The Elements of Political Economy* (New York, 1888), p. 349. A full-bodied account of the state of economic opinion in the United States during the 1870's and 1880's is given in Dorfman, *Thorstein Veblen, passim.*
8. " The Past and the Present of Political Economy," *Johns Hopkins University Studies in Historical and Political Science,* II (1884), 64, *passim.*
9. *The Premises of Political Economy,* pp. 78–79. John Bates Clark, at that early time in his career, was also sharply critical of classical economics; see *The Philosophy of Wealth,* esp. pp. iii, 32–35, 38 ff., 48, 65–67, 120, 147, 150, 186–96, 207.
10. Quoted in Ely, *Ground Under Our Feet* (New York, 1938), p. 140. Compare this with Ely's original draft, p. 136. For Ely's account of the Association see pp. 121–64. See also *Publications,* American Economic Association, I (1886), 5–36.
11. For a rather equivocal statement on competition, see Ely, " Competition: Its Nature, Its Permanency, and Its Beneficence," *Publications,* American Economic Association, Third Series, II (1901), 55–70. Dorfman emphasizes the essential conservatism of the " New School " leaders, *op. cit.,* pp. 61–64.
12. *Ground Under Our Feet,* p. 154. Cf. " The Past and the Present of Political Economy," *Johns Hopkins University Studies in Historical and Political Science* II (1884), 45 ff. See also F. A. Walker, " Recent Progress of Political Economy in the United States," *Publications,* American Economic Association, IV (1889), 31–32.
13. " Mr. Bellamy and the New Nationalist Party," *Atlantic Monthly,* LXV (1890), 261–62. Elsewhere, however, Walker asserted that the solidarity of

the family prevents the survival of the fittest from operating among men; *Political Economy*, pp. 300–1. For other uses of the struggle for existence, see Arthur T. Hadley, *Economics* (New York, 1896), pp. 19–22; John B. Clark, *Essentials of Economic Theory*, p. 274. A critical attitude was expressed by Herbert J. Davenport, *The Economics of Enterprise* (New York, 1913), pp. 20–21.

14. *The Theory of Dynamic Economics* (Philadelphia, 1892), chaps. i–viii, esp. pp. 18, 21, 24, 37–38. Patten's interest in consumption as a source of economic change was stimulated by the subjective approach of marginal-utility theory; see pp. 37–38. Cf. *The Consumption of Wealth* (Philadelphia, 1889).

15. *The Theory of Social Forces*, esp. pp. 5–17, 22–24, 52–53, 76–90.

16. See Rexford G. Tugwell, "Notes on the Life and Work of Simon Nelson Patten," *Journal of Political Economy*, XXXI (1923), 153–208; Scott Nearing, *Educational Frontiers, a Book about Simon Nelson Patten and Other Teachers* (New York, 1925).

17. See his most popular book, *The New Basis of Civilization* (New York, 1907).

18. *The Religion Worth Having*, passim.

19. See *Essays in Social Justice*, pp. 18, 19, 91–98, 103–4, 259.

20. *Journal of Political Economy*, V (1897), 99; *ibid.*, XI (1903), 655–56. On the relation between Ward and Veblen, see Dorfman, *op. cit.*, pp. 194–96, 210–11.

21. *The Theory of the Leisure Class* (New York, Modern Library, 1934), pp. 237–38.

22. *Ibid.*, chaps. viii–x. Veblen's treatment of business enterprise is much less severe in *The Theory of Business Enterprise* than in *Absentee Ownership*, esp. chaps. iii–vi. See also *The Engineers and the Price System* (New York, 1921).

23. *Journal of Political Economy*, VI (1898), 430–35.

24. *The Theory of the Leisure Class*, chaps. viii, ix, esp. pp. 188–91, 236–41.

25. "Why Is Economics Not an Evolutionary Science?" *Quarterly Journal of Economics*, XII (1898), 373–97. Cf. *The Theory of Business Enterprise*, pp. 363–65.

Particularly vulnerable to Veblen's approach was the later economics of John Bates Clark, whose *Essentials of Economic Theory* was a perfect example of the conception of things in which free competition was held to be a feature of "natural law." See "Professor Clark's Economics," *Quarterly Journal of Economics*, XXII (1908), 155–60. Veblen also looked upon Karl Marx's social theory as pre-Darwinian, although its essential preconception, being that of "the Hegelian Left" which assumed an inherent tendency to events to work toward a stated goal, was superficially different from that of classical economics. The simplicity of the Marxian notion of conscious class·struggle, Veblen believed, arose out of its affiliation with hedonism, whose defects it shared; the underlying notion of the "normal case" of class solidarity in pursuit of individual interest was closely akin to utilitarianism. See "The Socialist Economics of Karl Marx and His Followers," *ibid.*, XX (1906), 409–30, esp. 411–18. Veblen criticized the members of the historical school as falling short of modern science in having "contented themselves with an enumeration of data and a narrative account of industrial development" and for not having presumed "to offer a theory of anything or to elaborate their results into a consistent body of knowledge." *Ibid.*, XII, 373. See also "Gustave Schmoller's Economics," *ibid.*, XVI (1901), 253–55. The essays referred to in this note are collected in *The Place of Science in Modern Civilization and Other Essays* (New York, 1919).

26. "The Preconceptions of Economic Science," Part II, *Quarterly Journal of Economics*, XIII (1898), 425.

27. John M. Clark, "Problems of Economic Theory — Discussion," *American Economic Review*, XV, Supplement (1925), 56.

28. Keller, *Societal Evolution* (New York, 1915), p. 326; cf. pp. 250 ff.

29. "The Concepts and Methods of Sociology," *American Journal of Sociology*, X (1904), 172; *Studies in the Theory of Human Society* (New York, 1922), 136–41.

30. *Principles of Sociology*, p. v.

31. *The Responsible State*, p. 107; *Studies in the Theory of Human Society*, pp. 16–17, 206–7, 226, and chap. xiv; *The Elements of Sociology*, pp. 234–35, 293–95; *Inductive Sociology*, p. 6.

32. "The Persistence of Competition," *Political Science Quarterly*, II (1887), 66.

33. *The Responsible State*, p. 108; *The Elements of Sociology*, p. 317.

34. "The Principles of Sociology," *American Journal of Sociology*, II (1897), 741–42.

35. "The Failure of Biologic Sociology," *Annals of the American Academy of Political and Social Science*, IV (1894), 68–69. Patten's essay was, however, sadly misdirected, since it took to task none other than Ward for having fostered biological sociology.

36. *Democracy and Empire* (New York, 1900), p. 29.

37. *General Sociology*, p. ix.

38. "Fifty Years of Sociology in the United States," *American Journal of Sociology*, XXI (1916), 773.

39. "The Influence of Darwin on Sociology," *Psychological Review*, N. S., XVI (1909), 189.

40. *Darwin and the Humanities*, p. 40. See also *Social and Ethical Interpretations in Mental Development* (New York, 1897), pp. 520–23.

41. See Dewey, "The Need for Social Psychology," *Psychological Review*, XXIV (1917); Cooley, *Human Nature and the Social Order*, esp. chap. i. Cooley acknowledged the guidance of William James and Baldwin (p. 90 n.). Cf. also Cooley, *Social Organization*, chap. i; Baldwin, *Social and Ethical Interpretations in Mental Development*, pp. 87–88; *The Individual and Society*, chap. i. Many writers were considerably influenced by French social psychology, particularly by Tarde. For an analysis of the old psychology and the new trend, see Fay Berger Karpf, *American Social Psychology*, pp. 25–40, 176–95, 216–45, 269–307, 327–50.

42. *The Individual and Society*, p. 118.

43. *Human Nature and the Social Order*, p. 12.

44. *Human Nature and Conduct*, pp. 21–22.

45. This was not universally true. E. A. Ross used McDougall's instinct theory without deserting his earlier views. See *Principles of Sociology* (New York, 1921), pp. 42–43.

46. Cf. Cooley, *Social Organization*, pp. 120, 258–61, 291–96; *Social Process*, pp. 226–31.

47. Ross reported that his twenty-four books had sold over 300,000 copies. *Seventy Years of It*, pp. 95, 299.
48. *Ibid.*, p. 180.
49. *Principles of Sociology*, pp. 108–9; *Foundations of Sociology*, pp. 341–43; *Sin and Society* (New York, 1907), p. 53; *Seventy Years of It*, p. 55.
50. See *The Jukes* (New York, 1877), pp. 26, 39.
51. *American Charities*, chaps. iii–v.
52. For a typical alarmist view of the period, see W. Duncan McKim, *Heredity and Human Progress* (New York, 1899). For a moderate statement of the environmentalist viewpoint, see John R. Commons, "Natural Selection, Social Selection, and Heredity," *Arena*, XVIII (1897), 90–97.
53. *Proceedings of the First National Conference on Race Betterment* (Battle Creek, 1914).
54. For a review of the progress of eugenics legislation, see H. H. Laughlin, *Eugenical Sterilization: 1926* (New Haven, 1926), pp. 10–18.
55. See John Denison, "The Survival of the American Type," *Atlantic Monthly*, XXXV (1895), 16–28. See Charles B. Davenport in *Eugenics: Twelve University Lectures* (New York, 1914), p. 11.
56. See Paul Popenoe and Roswell H. Johnson, *Applied Eugenics*, chap. ii.
57. *The Social Direction of Human Evolution*, p. 44. For a criticism of this tendency in eugenics theory, see the excellent article by G. Spiller, "Darwinism and Sociology," *Sociological Review*, VII (1914), 232–53; also Clarence M. Case, "Eugenics as a Social Philosophy," *Journal of Applied Sociology*, VII (1922), 1–12.
58. Karl Pearson, Galton's successor in the international leadership of eugenics, in his little volume *National Life from the Standpoint of Science* (London, 1901), expressed a social philosophy as harsh as the worst effusions of German militarists.
59. *Hereditary Genius* (rev. Amer. ed., New York, 1871), pp. 14, 38–39, 41, 49.
60. Cited by Harvey E. Jordan in *Eugenics: Twelve University Lectures*, p. 110.
61. *The Kallikak Family* (New York, 1911), p. 116.
62. *The Heredity of Richard Roe* (Boston, 1911), p. 35.
63. "The Importance of the Eugenics Movement and its Relation to Social Hygiene," *Journal of the American Medical Association*, LIV (1910), 2018.
64. "Influence of Heredity on Human Society," *Annals of the American Academy of Political and Social Science*, XXXIV (1909), 16, 21. Cf. Davenport, *Heredity in Relation to Eugenics*, pp. 254–55; Edwin G. Conklin, *Heredity and Environment in the Development of Men* (Princeton, 1915), p. 206.
65. "Eugenics: with Special Reference to Intellect and Character," *Popular Science Monthly*, LXXXIII (1913), 128. Cf. Thorndike's *Educational Psychology* (New York, 1914), III, 310 ff.
66. "Eugenics," *Popular Science Monthly*, LXXXIII (1913), 134.
67. For the place of heredity in Thorndike's educational philosophy, see Curti, *The Social Ideas of American Educators*, pp. 473 ff. See also Thorndike's review of Lester Ward's *Applied Sociology*, "A Sociologist's Theory of Education," *Bookman*, XXIV (1906), 290–94.
68. Popenoe and Johnson, *op. cit.*, chap. xviii, "The Eugenic Aspect of Some Specific Reforms."

69. Alleyne Ireland, "Democracy and the Accepted Facts of Heredity," *Journal of Heredity*, IX (1918), 339–42; O. F. Cook and Robert C. Cook, "Biology and Government," *ibid.*, X (1919), 250–53; E. G. Conklin, "Heredity and Democracy, a Reply to Alleyne Ireland," *ibid.*, X (1919), 161–63. Popenoe and Johnson thought that the facts of biology call for an "aristo-democracy" which would preserve democratic parliamentarianism but give scope to the skill and training of specialists. *Op. cit.*, pp. 360–62.

70. Frederick A. Woods, "Kaiserism and Heredity," *Journal of Heredity*, IX (1918), 353.

71. *Applied Sociology, passim.* For his data Ward relied chiefly upon Alfred Odin's *Genèse des Grandes Hommes* (Paris, 1895), a study of environmental factors in the careers of over six thousand French men of letters.

72. Cooley to Ward, April 28, 1898, Ward MSS, Autograph Letters, VII, 8.

73. "Genius, Fame, and the Comparison of Races," *Annals of the American Academy of Political and Social Science*, IX (1897), 317–58.

74. Keller, *op. cit.*, pp. 193 ff.

75. *Social Process*, p. 206.

CHAPTER 9

RACISM AND IMPERIALISM

1. The fact that Darwin himself was not an unequivocal social Darwinist did not affect the plausibility of such appeals. For a discussion of Darwin's share in the responsibility for social Darwinism, see Bernhard J. Stern, *Science and Society*, VI (1942), 75–78.

2. See Jacques Novicow, *La Critique du Darwinisme Social* (Paris, 1910), pp. 12–15. The contributions of Darwinism to militarism and imperialism in European culture have been discussed in Carlton J. H. Hayes, *A Generation of Materialism*, pp. 12–13, 246, 255 ff., and in Jacques Barzun, *Darwin, Marx, Wagner, passim.*

3. "The Philosophy and Morals of War," *North American Review*, CLXIX (1889), 794.

4. Quoted by Julius W. Pratt in "The Ideology of American Expansion," *Essays in Honor of William E. Dodd* (Chicago, 1935), p. 344.

5. Albert J. Weinberg, *Manifest Destiny*, chap. vii.

6. See W. Stull Holt, "The Idea of Scientific History in America," *Journal of the History of Ideas*, I (1940), 352–62.

7. *Comparative Politics* (New York, 1874), p. 23.

8. *The Letters of Henry Adams*, II, 532.

9. Edward Saveth, "Race and Nationalism in American Historiography: The Late Nineteenth Century," *Political Science Quarterly*, LXIV (1939), 421–41.

10. *A Short History of Anglo-Saxon Freedom* (New York, 1890), p. 308.

11. *Ibid.*, p. 309.

12. *Political Science and Comparative Constitutional Law*, I, vi, 3–4, 39, 44–45.

13. See the preface by A. B. Hart, in *The Works of Theodore Roosevelt*, VIII, xiv.

14. *Ibid.*, VIII, 3–4, 7. While Roosevelt became aware of the vacuity of many of the common "racial" terms (such as "Aryan," "Teuton," and

"Anglo-Saxon ") , he was unable to shake off the incubus of racism. See *ibid.*, XII, 40–41, and his review of Houston Stewart Chamberlain, *ibid.*, 106–12. In 1896 he endorsed the racism of Le Bon as "very fine and true." *Selections from the Correspondence of Theodore Roosevelt and Henry Cabot Lodge* (New York, 1925) , I, 218.

15. *Outlines of Cosmic Philosophy,* II, 256 ff.
16. *Ibid.,* II, 263. See also *The Destiny of Man,* pp. 85 ff.
17. *Outlines of Cosmic Philosophy,* II, 341.
18. See *Civil Government in the United States,* p. xiii; *American Political Ideas, passim.*
19. *A Century of Science,* p. 222; *American Political Ideas,* pp. 43–44.
20. *American Political Ideas,* p. 135.
21. *Ibid.,* pp. 140–45.
22. Clark, *Life and Letters of John Fiske,* II, 139–40.
23. *American Political Ideas,* p. 7.
24. Clark, *op. cit.,* II, 165–67.
25. *Our Country,* p. 168.
26. *Ibid.,* p. 170, quoting *The Descent of Man,* ed. unspecified, Part I, p. 142; Darwin was referring to Zincke's *Last Winter in the United States* (London, 1868) , p. 29.
27. Strong, *op. cit.,* pp. 174–75. Cf. also Strong's *The New Era* (New York, 1893) , chap. iv.
28. Claude Bowers, *Beveridge and the Progressive Era* (Cambridge, 1932) , p. 121.
29. Roosevelt, *op. cit.,* XIII, 322–23, 331.
30. Tyler Dennett, *John Hay* (New York, 1933) , p. 278.
31. John R. Dos Passos, *The Anglo-Saxon Century,* p. 4.
32. "The Problem of the Philippines," *North American Review,* CLXVII (1898) , 267.
33. "The Economic Basis of Imperialism," *ibid.,* CLXVII (1898) , 326.
34. "Can New Openings Be Found for Capital?" *Atlantic Monthly,* LXXXIV (1899) , 600–8.
35. A. Lawrence Lowell, "The Colonial Expansion of the United States," *ibid.,* LXXXIII (1899) , 145–54.
36. George Burton Adams, "A Century of Anglo-Saxon Expansion," *Atlantic Monthly,* LXXIX (1897) , 528–38; John R. Dos Passos, *op. cit.,* p. x; Charles A. Gardiner, *The Proposed Anglo-Saxon Alliance* (New York, 1898) , p. 26; Lyman Abbott, "The Basis of an Anglo-American Understanding," *North American Review,* CLXVI (1898) , 513–21; John R. Procter, "Isolation or Imperialism," *Forum,* XXV (1898) , 14–26.
37. Hosmer, *op. cit.,* chap. xx.
38. Schurz, "The Anglo-American Friendship," *Atlantic Monthly,* LXXXII (1898) , 436.
39. *Atlantic Monthly,* LXXI (1898) , 577–88. Cf. Dos Passos, *op. cit.,* p. 57.
40. See *The Interest of America in Sea Power,* pp. 27, 107–34.
41. See Dennett, *op. cit.,* pp. 189, 219; Dos Passos, *op. cit.,* pp. 212–19, *passim; Selections from the Correspondence of Theodore Roosevelt and Henry Cabot Lodge,* I, 446; *An American Response to Expressions of English Sympathy;* Charles Waldstein, *The Expansion of Western Ideals and the World's Peace* (New York and London, 1899) .
42. Waldstein, *op. cit.,* pp. 20, 22 ff.

43. William R. Thayer, *Life and Letters of John Hay* (Boston, 1915), II, 234.

44. The best discussion is George L. Beer, *The English Speaking Peoples* (New York, 1917).

45. Stephen B. Luce, "The Benefits of War," *North American Review,* CLIII (1891), 677.

46. See Merle Curti, *Peace or War,* pp. 118–21; Harriet Bradbury, "War as a Necessity of Evolution," *Arena,* XXI (1891), 95–96; Charles Morris, "War as a Factor in Civilization," *Popular Science Monthly,* XLVII (1895), 823–24; N. S. Shaler, "The Natural History of Warfare," *North American Review,* CLXII (1896), 328–40.

47. Mahan, *op. cit.,* p. 267.

48. *National Life and Character,* p. 85.

49. *Letters,* II, 46.

50. *The Law of Civilization and Decay,* pp. viii ff.

51. *America's Economic Supremacy,* p. 192.

52. *Ibid.,* pp. 193–222.

53. "The New Industrial Revolution," *Atlantic Monthly,* LXXXVII (1901), 165.

54. Mahan, *op. cit.,* p. 18.

55. "National Life and Character," *op. cit.,* XIII, 220–22; "The Law of Civilization and Decay," *ibid.,* XIII, 242–60.

56. "Race Decadence" (1914), *op. cit.,* XII, 184–96. Cf. "A Letter from President Roosevelt on Race Suicide," [American] *Review of Reviews,* XXXV (1907), 550–57.

57. See J. F. Abbott, *Japanese Expansion and American Policies* (New York, 1916), chap. i.

58. Payson J. Treat, *Japan and the United States* (rev. ed., Stanford, 1928), p. 187.

59. For the outlook of a West Coast writer, see Montaville Flowers, *The Japanese Conquest of American Opinion* (New York, 1917).

60. Sidney L. Gulick, *America and the Orient* (New York, 1916), pp. 1–27.

61. "The Yellow Peril," in *Revolution and Other Essays* (New York, 1910), pp. 282–83.

62. "The Real Yellow Peril," *North American Review,* CLXXXVI (1907), 375–83. Cf. a more moderate view, J. O. P. Bland, "The Real Yellow Peril," *Atlantic Monthly,* III (1913), 734–44.

63. Abbott, *op. cit.;* S. L. Gulick, *The American Japanese Problem* (New York, 1914), chaps. xii, xiii. For samples of post-war alarmism see Madison Grant, *The Passing of the Great Race;* George Brandes, "The Passing of the White Race," *Forum,* LXV (1921), 254–56. Lothrop Stoddard feared that the doctrine of the survival of the fittest was beginning to prove a boomerang for the western peoples. See *The Rising Tide of Color,* pp. 23, 150, 167, 181–82, 219–21, 307–8.

64. *The Valor of Ignorance,* pp. 8, 11.

65. *Ibid.,* p. 44; cf. p. 76.

66. *The Day of the Saxon, passim.*

67. *Defenseless America,* pp. v, 27–41, 240.

68. See especially the foreword by Henry A. Wise Wood to W. H. Hobb's *Leonard Wood* (New York, 1920).

69. *Proceedings,* Congress of Constructive Patriotism, National Security League (New York, 1917), p. 16.

70. Hermann Hagedorn, *Leonard Wood* (New York, 1931), II, 173.

71. *Congressional Record,* 55th Congress, 3rd Session, p. 1424.

72. "Human Faculty as Determined by Race," *Proceedings,* American Association for the Advancement of Science, XLIII (1894), 301–27.

73. See James M. Baldwin, *Mental Development in the Child and in the Race* (New York, 1895), chap. i.

74. See *Adolescence* (New York, 1905), Vol II, chap. xviii, esp. pp. 647, 651, 698–700, 714, 716–18, 748.

75. See *Swords and Ploughshares* (New York, 1902), p. 54, *passim.*

76. Perry, *Thought and Character of William James,* II, 311.

77. *Arena,* XXII, 702.

78. Merle Curti, *op. cit.,* pp. 178–82. For representative anti-imperialist arguments, see David Starr Jordan, *Imperial Democracy* (New York, 1899); R. F. Pettigrew, *The Course of Empire* (New York, 1920), a reprint of speeches delivered in the Senate; George F. Hoar, *Autobiography of Seventy Years* (New York, 1903), Vol. II, chap. xxxiii. See also Fred Harrington, "Literary Aspects of American Anti-Imperialism," *New England Quarterly,* X (1937), 650–67. For left-wing arguments, see Morrison I. Swift, *Imperialism and Liberty* (Los Angeles, 1899).

79. Perry, *op. cit.,* II, 311.

80. Quoted, *ibid.,* II, 311–12.

81. "The Conquest of the United States by Spain," in *War and Other Essays,* p. 334.

82. See *The Blood of the Nation* (Boston, 1899); *The Human Harvest* (Boston, 1907); *War and Waste* (New York, 1912), chap. i; *War's Aftermath* (New York, 1914); *War and the Breed* (Boston, 1915).

83. See Theodore Roosevelt, "Twisted Eugenics," *op. cit.,* XII, 197–207; Hudson Maxim, *op. cit.,* 7–18; Charmian London, *The Book of Jack London,* II, 347–48.

84. "The New Internationalism," *Saturday Evening Post,* CXCIV (August 20, 1921), 20.

85. See William Archer, "Fighting a Philosopher," *North American Review,* CCI (1915), 30–44. "In a very real sense it is the philosophy of Nietzsche that we are fighting."

86. "The Lust of Empire," *Nation,* XCIX (1914), 493.

87. Quoted in *Out of Their Own Mouths* (New York, 1917), pp. 75–76.

88. Quoted, *ibid.,* p. 151.

89. *Germany vs. Civilization* (New York, 1916), pp. 80–81; *Volleys from a Non-Combatant* (New York, 1919), p. 20; cf. his preface to *Out of Their Own Mouths,* p. xv. See also Michael A. Morrison, *Sidelights on Germany* (New York, 1918), pp. 34 ff. For an English view, see J. H. Muirhead, *German Philosophy in Relation to the War* (London, 1915). An interesting contemporary defense of Germany is Max Eastman's *Understanding Germany* (New York, 1916), esp. pp. 60 ff.

90. "Blaming Nietzsche for It All," *Literary Digest,* XLIX (1914), 743–44; "Did Nietzsche Cause the War?" *Educational Review,* XLVIII (1914), 353–57.

91. Archer, *op. cit.,* pp. 30–31.

92. J. Edward Mercer, "Nietzsche and Darwinism," *Nineteenth Century*, LXXVII (1915), 421–31.
93. See G. Stanley Hall as quoted by Frederick Whitridge, *One American's Opinion of the European War* (New York, 1914), pp. 37–39. See also Hall's *Morale* (New York, 1920), pp. 10–14.
94. *The Present Conflict of Ideals*, pp. 425–28.
95. *Ibid.*, p. 145.
96. Curti, *op. cit.*, pp. 119–21.
97. See Novicow, *op. cit.*, and *Les Luttes entre Sociétés Humaines* (Paris, 1893).
98. *Social Progress and the Darwinian Theory*, pp. 21, 29, 53–60, 64–68, 79, *passim*.
99. *Ibid.*, p. 115.
100. *Headquarters Nights* (Boston, 1917).
101. See Wayne C. Williams, *William Jennings Bryan* (New York, 1936), p. 449.
102. *Seventy Years of It*, p. 88. Cf. Bryan's *In His Image* (New York, 1922), pp. 107–10, 123–26.

CHAPTER 10
CONCLUSION

1. Metaphorical appeals to biology have been commonly used in all ages, not only in the post-Darwinian period. Martin Luther in his discourse " On Trading and Usury " (1524) complained of large monopolies in these terms: " [They] oppress and ruin all the small merchants, as the pike the little fish in the water, just as though they were lords over God's creatures and free from all the laws of faith and love." And Falstaff argued, " If the young dace be a bait for the old pike, I see no reason in the law of nature but I may snap at him " (*Henry IV*, Part II, Act III, scene 2). Such examples could be multiplied indefinitely.

Index

Abbott, Francis Ellingwood, 22
Abbott, Lyman, 29, 106, 108
Adams, Brooks, 186–189
Adams, Henry, 15, 85, 173, 186
Adams, Herbert Baxter, 173–174
Agassiz, Louis, 17–18, 23, 26, 28, 127
Aguinaldo, Emilio, 194
American Association for the Advancement of Science, 18–19, 193
American Breeders' Association, 162
American Economic Association, 34, 107, 144, 147
American Genetic Association, 162
American Journal of Sciences and Arts, 13
American Journal of Sociology, 101
American Philosophical Society, 4
American Social Science Association, 47
American Sociological Society, 70, 82, 156
Anarchism, 7, 35, 59, 130, 134
Angell, Norman, 199
Anglo-Saxonism, 172–184, 191–194, 203
Anthropological Society of Washington, 72
Anthropologists, 4, 65, 85, 152, 169, 192
Anti-Corn Law League, 35
Anti-imperialism, 134, 183, 192, 194
Anti-Imperialist League, 194
Appleton's Journal, 22, 89
Aquinas, Saint Thomas, 26
Aristocracy, 9, 56, 71, 80, 82–83, 154, 157
Aristotle, 26, 31–32
Arnold, Matthew, 21
Atheism, 13, 18, 25–26, 30, 88
Atlantic Monthly, 23, 33, 87, 132

Baer, Karl Ernst von, 35
Bagehot, Walter, 23, 67, 90, 92, 112, 132

Bain, Alexander, 15, 23
Baldwin, James Mark, 158–159
Bancroft, George, 32, 177–178
Barker, Lewellys F., 164
Barnard, F. A. P., 31
Barrett, John, 181
Beard, Charles A., 60, 168–169
Beecher, Henry Ward, 29–31, 48
Behrends, A. J. F., 106, 108
Bellamy, Edward, 98, 107, 110, 113–114, 148
Bemis, Edward, 107
Bentham, Jeremy, 40–41
Bernhardi, Friedrich von, 190, 196–198
Beveridge, Albert T., 179–180, 182
Bible, 14, 16, 25–26, 28, 31
Biddle, Nicholas, 9
Bliss, William D. P., 106
Boas, Franz, 168–169, 192
Boehmert, Victor, 72
Bowen, Francis, 88, 146
Bowne, Borden P., 33
Brace, Charles Loring, 16, 22
Brandeis, Louis D., 121, 168–169
Brooks, Phillips, 30
Brown University, 69
Browning, Robert, 21
Brownson, Orestes A., 26
Bryan, William Jennings, 82, 119, 200
Büchner, Edward, 26, 36
Buckle, Thomas H., 15
Burgess, John W., 174–175
Burke, Edmund, 8
Butler, Nicholas Murray, 50

Calvinism, 10, 51, 66
Carnegie, Andrew, 9, 45, 49, 60
Carver, Thomas Nixon, 148, 151, 153, 198
Catholicism, 24, 26, 30, 87, 178
Chambers, Robert, 14
Chicago, University of, 82, 135
Christian Union, 29

Church Association for the Advancement of the Interests of Labor, 106
Churches: and evolution, 24–30, 105–110, 151
Clarke, James Freeman, 14
Clergy, 105–109. *See also* Churches; Social gospel
Cleveland, Stephen Grover, 183
Coleridge, Samuel T., 35
Columbia University, 157, 175
Commons, John R., 34, 107, 168
Competition, 6, 9, 35, 45, 52, 54, 57–59, 73–75, 86, 89, 93, 98–104, 108–110, 113–121, 140, 143–157, 164, 176–177, 185, 188, 201–204
Comte, Auguste, 20, 67, 82–83, 115, 137
Conant, Charles A., 181
Conservation of energy, 36, 127, 157
Conservatism, 5–10, 28, 41, 46–47, 51, 57, 80, 84, 88, 108, 110–111, 118, 121, 124, 136, 141, 157, 167
Contemporary Review, 55
Cooley, Charles H., 33, 143, 159–160, 166–167
Copernicus, Nicolaus, 3
Creel Committee on Public Information, 197
Croly, Herbert, 105, 121, 141
Crosby, Ernest Howard, 194
Cuvier, Georges L., 14, 17

Dana, James Dwight, 18, 29
Daniel, John W., 192
Darrow, Clarence, 34
Dartmouth College, 21
Darwin, Charles, 4–5, 13, 16–32, 36–39, 45, 55–56, 67, 77–79, 85, 88, 90–93, 109, 117, 132, 144, 147, 161, 167, 171, 178–179, 196–200
Darwinism, 4, 14–19, 23–24, 28, 85, 124–125, 136, 147–148, 154–155, 177; and psychology, 131–132, 150, 159; and theism, 25–31, 45, 86, 88, 108, 151; as new approach to nature, 3; social, 5–9, 11, 38, 43–44, 51–66, 68, 77, 81–82, 90, 95, 101, 104, 111, 137, 144, 152, 156–164, 170–172, 192, 196–203. *See also* Evolution; Natural selection
Davenport, Charles B., 164
Dawes, Henry L., 177
Dawn, 106
Debs, Eugene, 60
Democracy, 9, 56, 59–60, 63, 66, 71, 80, 82, 86, 100, 103, 119, 157, 166, 173–177, 188, 192–195, 200

Depew, Chauncey, 44
Descent of Man, The (Darwin), 24–27, 91–92, 171, 179, 200
Determinism, 51, 60, 68, 104, 125, 129–130, 157
DeVries, Hugo, 117, 163
Dewey, John, 33, 118, 123, 125, 134–142, 159–160, 168–169
Dickens, Charles, 21
Dickinson, G. Lowes, 134
Draper, John W., 23
Dreiser, Theodore, 34
Drummond, Henry, 90, 96–97, 103–104, 110
Dugdale, Richard, 161

Economist, 35
Economists, 4, 6, 34, 159, 187; and evolution, 143–156, 198, 202–203
Education, 41, 62–63, 72, 76–77, 84, 90, 95, 116, 127, 129, 134–142, 165–166, 185, 192
Eggleston, George Cary, 89
Eliot, Charles William, 19, 50, 127
Ellwood, Charles A., 158
Ely, Richard T., 34, 70, 107, 146–147
Emerson, Ralph Waldo, 32
Engels, Friedrich, 115
Enlightenment, 65
Ethics: Christian, 38, 86–87, 106, 110, 143, 201; and economics, 10–11, 51, 54, 61, 98, 115; and evolution, 40, 43, 85–104, 134, 138–140, 143–144, 198, 204; political, 79; Protestant, 51–52; utilitarian, 40
Eugenics movement, 82, 161–167, 185, 196
Everett, Edward, 32
Evolution, 3–4, 6, 13–17, 20, 22, 34–38, 51, 58, 68, 127; and purpose, 81; definition of, 129; evidences of, 19, 55; gradual, 117, 125; optimistic implications of, 16, 39, 44, 47–48, 78, 86, 89, 130, 176, 198; social, 42–43, 59–60, 66, 71, 73, 98, 115, 118, 133, 157, 171, 182, 198; speculative, 4, 14. *See also* Darwinism; Ethics; Natural selection

Ferri, Enrico, 152
Fichte, J. G., 31
Fiske, John, 13–15, 19–22, 24, 31–32, 48, 90, 94–96, 103–104, 110, 126, 139, 142, 176–178, 186, 198
Forum, 49, 71, 77
Fourteenth Amendment, 46–47
Free trade, 35, 63, 72

Freeman, Edward Augustus, 172–173
French Academy of Science, 23
Freud, Sigmund, 3
Fundamentalism, 25, 200

Galaxy, 24, 89
Galton, Sir Francis, 161, 164, 166
Gardner, Augustus P., 50
Garland, Hamlin, 34
Gary, Elbert H., 50
Geoffroy St. Hilaire, Etienne, 14
George, Henry, 47, 98, 107, 110–113
Giddings, Franklin H., 33, 79, 157–158
Gilman, Daniel Coit, 21
Gladden, Washington, 15, 105–106, 108–109
Gobineau, Comte Arthur de, 171
Goddard, Henry, 164
Godkin, E. L., 23, 134
Goethe, Johann Wolfgang von, 14
Grant, Ulysses S., 34
Gray, Asa, 13–14, 18–19, 23–24, 32, 38
Green, John Richard, 174
Gronlund, Laurence, 114–116
Gumplowicz, Ludwig, 77–78

Haeckel, Ernst, 26, 36, 55, 109, 115, 193
Hale, Edward Everett, 32
Hall, G. Stanley, 193
Hamilton, Alexander, 9
Harriman, Mrs. E. H., 162
Harris, William T., 33, 124
Harvard University, 13–14, 19–20, 106, 126, 128, 141
Hay, John, 179–180, 183
Hayes, Rutherford B., 177
Hegel, G. W. F., 31–32, 124, 128, 139
Helmholtz, Hermann L. von, 36
Herron, George, 106, 109–110
Herschel, Sir William, 15
Hewitt, Abram S., 46
Hill, David Jayne, 50
Hill, James J., 45
Hoar, George F., 177
Hobbes, Thomas, 91–92, 95
Hobson, John A., 101
Hodge, Charles, 26
Hodgskin, Thomas, 35
Holmes, Oliver Wendell, 32, 121, 126, 168
Holt, Henry, 34
Hooker, Joseph Dalton, 16
Hoover, Herbert, 200
Hosmer, James K., 174, 182
Howells, William Dean, 89

Howison, George, 33
Humboldt, Alexander von, 15
Huxley, Thomas Henry, 13, 15, 21, 26, 55, 90, 95–97, 104, 115, 139
Hyndman, H. M., 112

Iconoclast, The, 69
Imperialism, 56, 87, 91, 170–200, 202–203
Independent, 27–28, 62, 106
Individualism, 6, 34, 46, 49–50, 68, 72, 79–83, 91, 102–108, 114–122, 125, 133–134, 140–141, 146, 151, 156–159, 164–165, 168, 201–203
Industrialism, 5, 19, 35, 42, 45, 49, 58, 60, 73, 78, 97, 106, 113, 120, 152, 154, 159, 176, 185, 188–189
Ingersoll, Robert, 31
Instrumentalism, 68, 125, 135–136, 139, 141. See also Dewey; Pragmatism
International Science Series, 23, 43, 92
International Socialist Review, 116
Interstate Commerce Act, 62, 70
Iowa College, 106
Ireland, Alleyne, 166
Irish Land League, 112

James, Henry, 18, 127, 134
James, William, 17, 20, 32–33, 84, 118, 123–138, 141, 159, 194, 198, 201. See also Pragmatism
Jevons, Stanley, 23
Johns Hopkins University, 21, 70, 135, 173–174
Johnson, Roswell Hill, 165
Jordan, David Starr, 164, 195–196
Joule, James P., 36
Journal of Heredity, 166
Journal of Speculative Philosophy, 124, 130

Kant, Immanuel, 31
Keller, Albert Galloway, 156–157, 167
Kellicott, William E., 163
Kellogg, Vernon, 199–200
Kelvin, William Thomson, Lord, 36
Kemble, John Mitchell, 172
Keynes, John Maynard, 10
Kidd, Benjamin, 90, 99–102, 137, 198, 202
Kingsley, Charles, 172
Kipling, Rudyard, 193–194
Knights of Labor, 46, 105
Kropotkin, Peter, 90–91, 97, 104, 117, 140, 199, 201

LaFollette, Robert M., 121
Laissez faire, 5, 35, 40, 46, 51, 65, 68, 72–75, 79, 82, 84, 101, 108, 120, 123, 138, 144–146, 164
Lamarck, J., 17, 35, 39, 77, 98, 116
Lassalle, Ferdinand, 115
Laughlin, J. Laurence, 146
Laveleye, Émile de, 108
Lea, Homer, 170, 190–191
Le Conte, Joseph, 17, 23, 28–29, 38
Lewes, George H., 15
Lewis, Arthur M., 115–117
Lippmann, Walter, 122, 141
Lloyd, Henry Demarest, 120
Locke, John, 137
Lodge, Henry Cabot, 50, 179, 183
London, Jack, 34, 189
Lowell, James Russell, 32
Lubbock, Sir John, 92
Luce, Stephen B., 184
Lusk, Hugh, H., 189
Lyell, Sir Charles, 14–16, 19, 26, 35, 85

McCosh, James, 27, 29, 33, 38, 86
McDougall, William, 160
Machiavelli, Niccolo, 87
Mahan, Alfred T., 183–184, 188
Mallock, William H., 90, 102, 104, 153
Malthus, Thomas Robert, 35, 38–39, 51, 55–56, 65, 78–79, 85, 88, 91, 110–111, 145–147
Marble, Manton, 24
Marsh, Othniel C., 19–20, 55
Martineau, Harriet, 52–53, 64
Marx, Karl, 115–116, 117
Marxism, 46, 83, 98, 115, 117
Materialism, 26, 172, 187
Maudsley, Henry, 23
Maxim, Hiram, 191
Maxim, Hudson, 191
Mayer, Julius R., 36
Mendel, Gregor, 163
Menken, S. Stanwood, 191
Mercer, J. Edward, 198
Middle class, 11, 35, 63–64, 119–120, 202
Militarism, 56, 192, 202–203; American, 170–172, 182, 184, 190–191, 194, 196; German, 171–172, 190, 196–200
Mill, John Stuart, 15, 21, 126
Mivart, St. George, 27
Moleschott, Jacob, 36
Moltke, Helmuth von, 172
Mongredien, Augustus, 72

Monism, 67–68, 81, 84, 115, 117–118, 127
Monist, 139
Moody, Dwight L., 25
More, Paul Elmer, 197
Morgan, John Pierpont, 9
Morris, George Sylvester, 135
Morse, Edward S., 19
Müller, Max, 173

Nasmyth, George, 199
Nation, 23–24, 35, 126, 134
National Conference on Race Betterment, 162
Nationalist, 115
Natural causes, 16, 132–133
Natural law, 6, 41, 61, 65–68, 72–75, 84, 96, 108–109, 127, 144–146, 152, 154
Natural rights, 8, 40, 55, 59, 65–66
Natural selection, 4, 15–18, 22–27, 36–39, 42, 44, 51, 57–58, 64, 75–79, 85–99, 109, 116, 118, 140, 144–145, 151, 160–161, 170, 176, 179, 198, 201, 203. *See also* Darwinism; Evolution
Naturalism, 15, 36, 70, 87, 107, 123–126, 171, 204
Nebraska, University of, 200
New Deal, 9, 119, 122, 141
New Englander, 28
New Nation, 114
New York *Tribune*, 24, 64
New York *World*, 24
Newton, Sir Isaac, 3, 23, 36, 145
Nietzsche, Friedrich, 38, 86, 196–198
Nineteenth Century, 198
Nordau, Max, 171
North American Review, 22, 45, 101, 124, 126, 171
Notestein, Wallace, 197
Novicow, Jacques, 199

Oken, Lorenz, 17
Olney, Richard, 183
Origin of Species (Darwin), 13, 18–19, 22, 28, 91, 115, 141, 158, 171
Ostwald, Wilhelm, 36
Outlook, 29

Paley, William, 25
Patten, Simon, 146–149, 150–151, 158
Peabody, Francis Greenwood, 106
Pearson, Charles, 185–186, 188
Pearson, Karl, 164
Peirce, Charles, 32, 125–128, 130, 132, 137

Perry, Arthur Latham, 146, 164
Perry, Ralph Barton, 198
Phelps, William Lyon, 54
Phillips, Wendell, 32
Pluralism, 128
Plutocracy, 9, 11, 63, 200
Political economy. *See* Economists
Popenoe, Paul, 165
Popular Science Monthly, 22-23, 43, 127
Populists, 9, 46, 82, 119, 160
Porter, Noah, 20-21, 53, 64
Positivism, 20, 27, 37, 87, 126
Pragmatism, 5, 32, 49, 68, 84, 104, 118, 123-142
Presbyterianism, 27, 34
Princeton Review, 26
Princeton University, 27
Progress, 6, 31, 40, 44, 47, 59, 61, 75, 78-80, 85-86, 93, 99-103, 111-112, 138, 149-150, 178, 180, 199
Protestantism, 24, 26, 30, 105, 110, 178
Psychiatry, 162
Puritanism, 29, 87

Quarterly Journal of Economics, 154

Racism, 170-200
Ratzenhofer, Gustav, 77
Rauschenbusch, Walter, 107-108
Reid, Whitelaw, 21
Ricardo, David, 51-52, 64-65, 78, 145, 150
Rice, W. N., 25
Ripley, George, 24
Ripley, William Z., 193
Rockefeller, John D., 45
Roosevelt, Franklin D., 9
Roosevelt, Theodore, 101-102, 121, 164, 170, 175, 179-180, 183-184, 188-189, 195
Root, Elihu, 50
Ross, Edward A., 70, 82, 156, 158, 160, 164, 200
Rousseau, Jean Jacques, 74
Royce, Josiah, 33, 128-129

Saturday Evening Post, 196
Say, Jean B., 145
Schelling, F. W. von, 31
Schurz, Carl, 48, 183
Scientists, and evolution, 14-23
Scopes, John T., 200
Secularism, 7-9, 30, 107
Shakespeare, William, 133
Shaler, Nathaniel S., 90

Sherman, William T., 177
Sherman Act, 70, 120
Silsbee, Edward, 13-14
Sinclair, Upton, 61, 63
Single Tax, 46, 107, 110
Small, Albion W., 33, 70, 84, 156, 158
Smith, Adam, 145
Smith, Goldwin, 87-88
Smith, J. Allen, 60
Social gospel, 105-110
Social reform, 7-8, 16, 40, 43, 46-47, 60-65, 68, 71, 75, 80, 84, 99-100, 105, 118-121, 125, 134, 142, 161, 165, 167, 202
Socialism, 43, 46, 49, 54, 57, 61-65, 80, 83, 99-110, 114, 118, 135, 140, 152, 165, 167, 185. *See also* Marx; Marxism
Socialist Labor Party, 114
Socialist Party, 106-107
Sociologists, and evolution, 4, 41, 43, 48-49, 55, 59, 67-69, 84, 101-102, 110, 117, 133, 137, 143, 156, 158, 160-161, 198-199
Solidarism, 104, 110, 202
Spanish-American War, 11, 64, 170, 183, 185, 195-196
Sparks, Jared, 32
Spencer, Herbert, 4-6, 11-15, 20-29, 31-51, 55-57, 60-61, 67-72, 77-86, 90-92, 98-105, 108-119, 122-135, 138, 142-145, 153, 156-161, 164, 176, 178, 182-186, 198-202
Stanford University, 195
Sterilization, 162
Stoll, Elmer E., 197
Stone, Harlan Fiske, 50
Strong, Josiah, 107, 178-179, 186
Sumner, Charles, 32-33
Sumner, Thomas, 52
Sumner, William Graham, 6-12, 20-21, 48, 51-66, 69-70, 79, 82, 85, 90, 109, 117, 119, 122, 143-148, 151-153, 156-157, 164, 195, 201, 203
Sun Yat-sen, 190
Supreme Court, 46
Swift, Morrison I., 130

Tennyson, Alfred, 21, 182
Thayer, William Roscoe, 197
Theism, 18, 25-31, 108, 151
Thorndike, Edward Lee, 165
Ticknor, George, 32
Tolstoi, Leo, 194
Transcendentalism, 32-33
Treitschke, Heinrich von, 196
Tufts, James H., 140

Turner, Frederick Jackson, 168–169, 178
Tylor, Edward, 23, 92, 173
Tyndall, John, 21, 23

Unitarianism, 33

Van Hise, Charles R., 121
Vanderbilt University, 21
Veblen, Thorstein, 65, 82, 143–145, 152–156, 159, 168–169
Vincent, George E., 158
Vogt, Karl, 12

Wagner, Klaus, 197
Waite, Morrison R., 177
Walker, Francis Amasa, 144–145, 148, 153
Wallace, Alfred R., 24, 39, 94, 144
Walling, William English, 117–118
Ward, Mrs. Humphrey, 101
Ward, Lester F., 5, 33, 67–85, 114–117, 120, 122, 137–138, 149, 152, 156, 159–160, 166, 201
Warner, Amos G., 162

Washington *National Union*, 71
Wayland, Francis, 53, 145
Weismann, August, 77, 98–99, 116, 163, 166
Wesleyan University, 25
Weyl, Walter, 120
Whitman, Walt, 14, 44
Whitney, William C., 53
Wilberforce, William, 13
Wilson, Woodrow, 50, 120–121, 151
Winchell, Alexander, 21
Wood, Leonard, 191
Woods, Frederick Adams, 166
Woolsey, Theodore Dwight, 53
World War I, 166, 190, 196–197, 200, 203
Wright, Chauncey, 22, 32, 125–126, 137
Wyman, Jeffries, 127

Yale University, 11, 19–20, 51, 53, 64, 152, 195
" Yellow Peril," 185, 189–190
Youmans, Edward Livingston, 14, 22–23, 27, 31–34, 47, 49, 92, 142

Printed in the United States
By Bookmasters